Break Up the Anthropocene

Forerunners: Ideas First

Short books of thought-in-process scholarship, where intense analysis, questioning, and speculation take the lead

FROM THE UNIVERSITY OF MINNESOTA PRESS

Steve Mentz
Break Up the Anthropocene

John Protevi
Edges of the State

Matthew J. Wolf-Meyer
Theory for the World to Come: Speculative Fiction and Apocalyptic Anthropology

Nicholas Tampio
Learning versus the Common Core

Kathryn Yusoff
A Billion Black Anthropocenes or None

Kenneth J. Saltman
The Swindle of Innovative Educational Finance

Ginger Nolan
The Neocolonialism of the Global Village

Joanna Zylinska
The End of Man: A Feminist Counterapocalypse

Robert Rosenberger
Callous Objects: Designs against the Homeless

William E. Connolly
Aspirational Fascism: The Struggle for Multifaceted Democracy under Trumpism

Chuck Rybak
UW Struggle: When a State Attacks Its University

Clare Birchall
Shareveillance: The Dangers of Openly Sharing and Covertly Collecting Data

la paperson
A Third University Is Possible

Kelly Oliver
Carceral Humanitarianism: Logics of Refugee Detention

P. David Marshall
The Celebrity Persona Pandemic

Davide Panagia
Ten Theses for an Aesthetics of Politics

David Golumbia
The Politics of Bitcoin: Software as Right-Wing Extremism

Sohail Daulatzai
Fifty Years of *The Battle of Algiers*: Past as Prologue

Gary Hall
The Uberfication of the University

Mark Jarzombek
Digital Stockholm Syndrome in the Post-Ontological Age

N. Adriana Knouf
How Noise Matters to Finance

Andrew Culp
Dark Deleuze

Akira Mizuta Lippit
Cinema without Reflection: Jacques Derrida's Echopoiesis and Narcissism Adrift

Sharon Sliwinski
Mandela's Dark Years: A Political Theory of Dreaming

Grant Farred
Martin Heidegger Saved My Life

Ian Bogost
The Geek's Chihuahua: Living with Apple

Shannon Mattern
Deep Mapping the Media City

Steven Shaviro
No Speed Limit: Three Essays on Accelerationism

Jussi Parikka
The Anthrobscene

Reinhold Martin
Mediators: Aesthetics, Politics, and the City

John Hartigan Jr.
Aesop's Anthropology: A Multispecies Approach

Break Up the Anthropocene

Steve Mentz

University of Minnesota Press
MINNEAPOLIS
LONDON

Published by the University of Minnesota Press, 2019
111 Third Avenue South, Suite 290
Minneapolis, MN 55401–2520
http://www.upress.umn.edu

Contents

Painting by Vanessa Daws. Dublin, 2018. Acrylic on canvas board. You can see a color image on my website: http://stevementz.com/pluralizetheanthropocene/.

Plural Ships on Plural Seas

THE REAL THING is round, painted, about eight inches in diameter, and hangs in a square frame in my house in Connecticut. Vanessa Daws, swim artist from Dublin, painted the image for me in June 2018.

Like the Anthropocene, this painting is something we can't all look at together but need to understand. I'm going to open this book by describing it to you.

The fragile ship balances atop plurality, toxicity, beauty. Multiplicities teem beneath its keel. The small circular painting blazes forth many Anthropocene conditions. We need more of these things in our breaking world: more ships, more seas, more monsters, more art.

Ship

Off center floats a ship, its solitary mast angling into dark skies. Blank faces crowd the deck, attracting and repelling attention. "Ship of fools": it's perched atop the trash swell, a toxic pile that cascades down and to the right. Unseen eyes lurk behind empty faces. Is that the Old Man himself up in the crow's nest pointing his spyglass at the horizon? The vessel holds the preterite ones, those who have been passed over, who soon may plunge into iridescent waters. Looking closer, the faces present different colors, textures, histories. Why can we see no eyes? Only Anthropos up high aims his glass, looking ahead into the circle's dark rim. He sails into futures we can't see.

Seas

The waters boil with trash and color. The space deep below blurs indefinitely, but near the surface it's possible to discern bottles, packaging, even a label on which is printed the letter "C." All of it is plastic and predetermined. The vessel traverses the Great Pacific Garbage Patch, currently three times the size of France, the most massive assemblage humans have yet produced. The image's tilt forces our eyes, with the ship, to lurch downward from left to right, escaping from a lurid tentacle of yellow-green froth on one side but heading toward more swells below. The ship provides partial respite, some possibility of story and change. But the circular frame tells its own tale: what happens when all waters everywhere throb with these same colors, these same plastics, this toxicity? None of the humans on board looks at the water with longing. I imagine a sign: No Swimming. This painting, made by a swim artist, frames inaccessible waters.

Monster

The creature's tail arcs behind, curling interrogatively back onto itself. The image copies one half of a monster that Rachel Carson reproduced for the endpapers of her 1951 masterpiece *The Sea around Us*. Carson took the sea snake from a map she found in the New York Public Library, *Il Mari di Amazones*.[1] The image was

1. Rachel Carson, *The Sea around Us* (New York: Oxford University Press, 1951), vi.

first engraved by the Italian master Antonio Francesco Lucini in 1646, when he was working on a massive atlas compiled by the renegade English nobleman and mariner Sir Robert Dudley.[2] What can the creature see beneath the ship, in purple-green semidarkness? Perhaps the monster trails the wake for the trash humans cast off the ship's stern. Perhaps the great curved tail will extend itself with a colorful splash onto the sea surface—whapp!—as the invisible body turns down into the depths. No one on the ship sees the monster's head. Even Anthropos on the masthead looks away, down the swell, into futurity. Monsters swim from oceans of the past, through currents of history and windstorms of legend. This one may paddle next to us for a long time to come.

Art

There's a communal joy in turning our eyes together onto a single image. The picture aligns us into community. We all look the same direction.

Like me, the artist who painted this picture is an open-water swimmer. Vanessa Daws and I first met, at least in the world beyond Facebook, in the California surf in October 2014. We were gathered together that week in Santa Barbara by the medieval maestros of the BABEL Working Group, which organized a rollicking seaside conference.[3] The Pacific was uncommonly gentle and warm. I remember seeing a harbor seal's head poking up from the waves next to me on my last morning swim. Looking

2. Dudley's atlas, *Dell'Arcano del Mare,* was published in Florence in 1645–46. My thanks to Chet van Duzer for helping me identify that Carson's map originally comes from Lucini and Dudley's volume.

3. On the BABEL-sponsored event in Santa Barbara, see my blog-wrapup: http://stevementz.com/babel-2016-preview-heading-to-the-beach-in-santa-barbara/. Accessed July 18, 2018.

at the painting today in my home on the Connecticut shoreline, I see angry water. The colors burn like garbage on fire, garish and toxic. There's truth in those swirling waters, the sort of truth that stings when you touch it.

The image recalls plural waters: the warm salt of Long Island Sound down the street, the surf break at Hendry's in Santa Barbara, Vanessa Daws's Irish Sea, the Clevedon Marine Lake in the West Country where she and I last swam together. Must they all churn yellow and green in our soon-arriving futures?

How everlasting are things in this Anthropocene in which our bodies are immersed today? How much color can flow into each solitary mind or splash across each swimming body? How can we love this fragile and partly inaccessible world, its flashes of light and its slow dripping melt?

Pluralize the Anthropocene!

THE EARTH BURNS HOT, but I've read the books.

The rapidly growing discourse of Anthropocene humanities challenges readers and thinkers to engage with a field in motion, in which new books, articles, and journals appear almost weekly. Rival 'cenes have been straining forward and puffing their chests. What we need from this seething cauldron of rival terms and points of view is not a discourse of mastery—one 'cene to rule them all—but a route into plurality. The Age of Anthropos invites introspection and political outrage, but the now-unavoidable 'cenic term also requires careful parsing. Faced with multiple narratives in multiple, overlapping discourses from the sciences to many kinds of humanities, all of which claim to describe our unstable environmental *now,* readers and scholars may be forgiven for a certain befuddled or baffled attitude. What are we meant to do with all this Anthropocene writing? To which question I proffer a #hashtag ready for bumper stickers and coffee cups: #pluralizetheanthropocene!

The pluralizing project emerges from an urgent need to renarrate the apocalyptic story in which Old Man Anthropos destroys the world. The material substrate of that narrative of doom seems perfectly accurate: the burning of fossil fuels has destabilized the earth system, with violent and unsettling consequences. Whether you call the unprecedented results "global warming," the Anthropocene, the "sixth extinction," or, in Bruno Latour's now-preferred term, the "New Climactic Regime," the contours of this disturbed world are legible in our cities, bodies, and systems of measurement.[1] We need to take stock and situate ourselves amid disorder. Familiar responses to immensity and change, from religious awe to the Romantic sublime, seem inadequate or inappropriate to the task at hand. In a compelling aside in *New Literary History,* Latour laments the loss of "the feeling of the sublime" that since classical antiquity described the heights of literary meaning, but he also anticipates a renovation or replacement of such elevation: "What's next? The successor of the sublime is under construction" (476). Rather than replace or remodel the Romantic superstructures we have inherited from the philosophical traditions of Kant and Burke and the literary forms of Milton and Shakespeare, it may be time to assemble something new, or at least newly dynamic. We must admit nonhuman disorder into the fold, as Latour and many others insist. But perhaps more importantly, we must also salvage amid our ruins the humane values of openness and sympathy. A plural Anthropocene seeks justice and embraces difference.

A pluralized Anthropocene proposes only partial orders. The pluralizing project refuses mastery and fantasies of wholeness in favor of a dynamic and disorderly system. An agglomeration of partial overlapping and sometimes conflicting perspectives replaces the singularity of capital-A Anthropos. We're left with messy assemblages overflowing with human and nonhuman agents and structures. This plural hodgepodge recovers from classical linguistics the gender plurality that both the Greek *an-*

1. Bruno Latour, "Life among Conceptual Characters," *New Literary History* 47, no. 2–3 (2016): 463–76 at 474.

thropos and the Latin *homo* obscure; these words mean "Man" in a collective sense, but under patriarchy they have a tendency to refer only to males. Among the key positive values of a pluralized Anthropocene are a flexible approach to scale, a capacity for dynamic speculative range, the ability to respond to catastrophic change, and self-reflexive curiosity. These capabilities, as some readers will have already noted, are often grouped among the special habits or talents of humanist thinking. That's good: we need the human and posthuman both. In what follows I aim to brew a rich combination of human and nonhuman flavors. To pluralize means to entangle and enter into: the Anthropocene that results from this process will be intimate and painful, resonating with stories from the human past and gesturing toward an uncertain future. To get to plurality we must break up the hard soil of the Anthropocene.

This project refuses any Anthropocene monoculture. On a basic level, that refusal includes rejecting the masculine and technocratic monopoly of the Anthropocene Working Group, which has recently settled on 1950 as its "Golden Spike" origin date, thus making the Anthropocene roughly synonymous with the Great Acceleration after World War II.[2] Beyond Golden Spike–ism, however, I set pluralization against any rigid analysis of the Anthropocene as a purely physical phenomenon, from the works of scientific thinkers such as Paul Crutzen, Eugene Stoermer, and Jay Zalasiewicz, to activist Bill McKibben's focus on the magic number 350, upon which he founded 350.org in 2008, with the number indicating the maximum allowable parts per million

2. Jan Zalasiewicz, C. P. Summerhayes, Colin Neil Waters, Mark Williams et al., "The Working Group on the Anthropocene: Summary of Evidence and Interim Recommendations," *Anthropocene* 19, no. 55–60 (September 2017): https://www.researchgate.net/publication/319613362_The_Working_Group_on_the_Anthropocene_Summary_of_evidence_and_interim_recommendations. Accessed July 19. 2018.

of carbon in the atmosphere.[3] McKibben and the scientists on whom he relies aren't wrong about the history of carbon—they are painfully, brutally accurate—but their nostalgia for a time "before" blinds them to the presence of Anthropocene effects prior to industrialization. I instead proffer a composting model of anthropogenic climate change, in which human activities have been destroying and remaking nonhuman environments since the earliest uses of fire, with massively more destructive accumulations of carbon since the Industrial Revolution layering themselves on top of a preindustrial base. Human history in this model represents the accumulation and thickening of anthropogenic pressures as human populations expand in numbers and geographic range. Refusing monocultures requires that we entangle multiple historical Anthropocenes, intermingling scientific observations and cultural obsessions. We must refuse the antihistoricism of once-and-for-all origin stories and the anthropocentrism that can only make sense of human stories on human scales.

To construct a pluralized Anthropocene, multiple possible conceptions of this epoch must overlap, connect, and entangle in ways that may seem only partially legible. A posthuman network of human and nonhuman actors may be difficult to reconcile with humane desires for racial, sexual, and class-based justice—but no one said the Anthropocene was going to be easy. The essential technology for engaging plurality, however, is one we already have: stories. The pluralizing project, which I take to be the essential contribution that humanities scholars can make to Anthropocene discourse, entails recognizing the controlling function of narrative in human history. Stories facilitate

3. For my earlier quibble with McKibben regarding the number 350, see "Tongues in the Storm: Shakespeare, Ecological Crisis, and the Resources of Genre," in *Ecocritical Shakespeare*, ed. Lynne Brucker and Dan Brayton, 155–71 (Aldershot, U.K.: Ashgate Publishing, 2011).

our responses to Anthropocene forces. Narrative cultures create mechanisms and patterns that engage, manipulate, and cherish radical change. The historical wisdom of humanities scholarship can help pluralize and make legible the swirling cauldron of discourses that populate our environmental present.

The project will require some steps along the way.

Untranslating Old Man Anthropos

His name is a problem. It's too male, too Greek, too monolithic. Contemporary theorists may argue that the Anthropocene is, counterintuitively, a nonanthropocentric moment, even, in Timothy Morton's breathless phrase, that "*'Anthropocene' is the first fully antianthropocentric concept*."[4] But even with this jiu-jitsu, Old Man Anthropos takes up too much space, in the world, in critical theory, and as the cause of so much ecological devastation. Anthropos occupies the dominant and destructive positions: old, rich, male, presumptively white, brutally normative. Before we can make progress with the term, we need to unweave the forces hidden within the figure.

Treating Anthropos as a human figure can help conceptually escape from his tyranny. If he's Man, he's also merely human. He stands and stoops and falls and complains. The word *anthropos* conceals, behind its singularity, the myriad of associations we have with the idea of the human, including humanism and humane ethical behavior. A key interpretive step has recently been proposed by Phillip John Usher as "Untranslating the Anthropocene."[5] Usher's philological project observes that the word *Anthropocene* has meaningful

4. Timothy Morton, *Dark Ecology: For a Logic of Future Coexistence* (New York: Columbia University Press, 2016), 24. Italics in original.

5. Phillip John Usher, "Untranslating the Anthropocene," *diacritics* 44, no. 3 (2016): 56–77.

parallels with the exterior analytical position imagined by the cognate term *anthropology*. The subject position of the purportedly objective observer, whether a climate scientist or an anthropologist doing field work, encourages a dissociation between anthropos and the self. In Usher's phrasing, "the very term 'Anthropocene' . . . sets up a certain relationship between *little me*, amateur or professional user of the term, and this *other* human *over there*" (60, italics in original). To separate the *me* who feels from the *other* who fouls the climate may salve our consciences, but that *me* needs its *other*. Usher's further speculations link the anthropos to the sciences and to an externalized conception of Nature, while the Latin word *homo* exfoliates into human, the humanities, and eventually to Culture (62). That provisional separation, as with Latour's famous description of modernity, turns out to be pure fantasy: "the word 'Anthropocene,'" Usher concludes, "already calls out for the rebellious cat that is the humanist" (70). That nonhuman cat, linked in Usher's analysis to both Montaigne's and Derrida's feline-inflected antiessentialisms, entangles humanist thinking with the Anthropocene in more-than-human figures and meanings. We don't just need the Old Man tallying up his stock receipts and reading the newspaper. We also need his hungry cat, eyeing the frail human flesh and only temporarily mollified by the kibble that appears every morning in the bowl.

Human, Posthuman, Alien, Ocean

The posthuman charges into Anthropocene discourses brandishing a liberator's torch, but this theoretical move risks losing contact with embodiment and human experience. One solution, or perhaps dis-solution, can be found in Astrida Neimanis's "posthuman feminist phenomenology" and her focus on humans as "bodies of water."[6] Starting from the physical reality that water composes most of our bodies and most of our planet's surface and biosphere, Neimanis develops a posthuman ethics that insists on responsibility for vulnerable bodies while also recognizing that water's circulation and dissolution always exceeds bodily borders. She highlights the "paradox of bodies . . . that we are willing to defend [them] to the death, even as we know they are falling apart at the seams" (18). Her analysis demixes human and feminist calls for justice with posthuman decentering. In "negotiating" rather than attempting to solve this paradox, Neimanis opts to "stay with the trouble," in Donna Haraway's phrase.[7] Human justice and posthuman disruption emerge entwined in her conception of embodied fluidity.

Neimanis's hydrofeminist approach enables the values of humanism to seep into the embodied complexity of the posthuman. The water that makes up most of our bodies comprises an internal other-ness, an alien substrate from and through which human selves connect to more-than-human environments. While valuing the posthuman gambit that unlocks the prison door of anthropocentrism, Neimanis insists "on our own situatedness as bodies that are *also still human*" (26, italics in original). To be human and ocean, self and alien, speaking subject and dynamic fluid, entails imagining across boundaries and into plural figurations. Neimanis takes inspiration from poststructural feminist thinkers such as Elizabeth Grosz and Stacy Alaimo as well as the phenomenology of Merleau-Ponty and the hydrophilic radicalism of Luce Irigaray's *Marine Lover (of Friedrich Nietzsche)*. Seeking "intimacy" rather than "mastery" (112), Neimanis combines oppositional forces so that fluid bodies assume particular

6. Astrida Neimanis, *Bodies of Water: Posthuman Feminist Phenomenology* (London: Bloomsbury, 2017).

7. Donna Haraway, *Staying with the Trouble: Making Kin in the Chthulucene* (Durham, N.C.: Duke University Press, 2016).

potency in an Anthropocene context. Attention to the complexities of local "bodies of water" operates as "an antidote to Anthropocene water" (171) as globalizing totality. Turning to the project of "learn[ing] to swim" (26) as a figure for encountering inhuman environmental complexity with merely human bodies, Neimanis challenges Anthropocene thinkers to juxtapose intimate human scales with world-sized ecological disasters.

Wages of Catastrophe

The crucial techniques for naming and isolating the Anthropocene have come from the geological sciences, as these discourses have since the nineteenth century developed a language and analytic framework to narrate the deep history of the earth. According to English professor Jeremy Davies, the crucial insight that engendered the Anthropocene was the rise of "neocatastrophism" as a geological paradigm over the past forty years. In his brilliant study *The Birth of the Anthropocene,*[8] Davies makes a deceptively simple claim: "the idea of the Anthropocene should be seen as another product of that neocatastrophist turn" (9). The seeming simplicity of the turn to catastrophes rather than gradualism in earth history, however, belies the radical implications of the shift. Well-established environmental concepts such as "sustainability" become untenable in a neocatastrophist context; in place of sustainability Davies enjoins us to be concerned "above all with environmental injustice and with fostering ecological pluralism and complexity" (6). Geologic time becomes not just anti-anthropocentrically vast but also dynamic, full of ruptures and unexpected turns. This Anthropocene requires not technical fixes but learning to

8. Jeremy Davies, *Birth of the Anthropocene* (Berkeley: University of California Press, 2018).

"live within the crisis" (194). In this disorienting space, human structures remain fragile but also deeply needed.

Davies makes a series of claims that flow from understanding the Anthropocene as an extension of neocatastrophist geological thinking (108–11). In placing catastrophe at the center of his model, he eschews a vast family of "green" eco-values such as the balance of nature. His Anthropocene isn't a broken harmony but a consequence of human-fueled accelerants driving forward a system that was never reliably stable to start. He asks that we radically de-center the human while also refusing any dualism that would separate humans from "nature." He sees human societies, not "Anthropos" as such, as crucial environmental actors. He argues that the "politically salient issue" is less the possible state of the planet at some future time but our current "time of transition into the Anthropocene" (110). Finally, he argues that the essential response to that transition must be not technical or philosophical but fundamentally political. In this last sense, Davies emphasizes what he sees as the true novelty of the new epoch: not questions of blame for human-driven climate disruptions but stark physical markers, including atmospheric carbon at "levels not seen for three million years" (110). The materiality of the Anthropocene turns out to be its most powerful feature.

The term *catastrophe* emerges from classical dramaturgy, as an English professor like Davies recognizes. A dramatic catastrophe describes the final turn (*strophi*) down or away from (*kato-*) the main action; it's the last plot twist before resolution. In the deep time of the neocatastrophic, however, no final resolution waits; each radical turn represents both destruction and possibly a new beginning, but in an utterly unfamiliar context. As ecotheorist Timothy Morton remarks, the most catastrophic era in the fossil record may be the Great Oxygenation Event that occurred between two and three billion years ago and wiped out almost all anaerobic life (*Dark Ecology*, 70). Geological eras,

Morton observes, are "*nested* catastrophes" (70) that also represent, playing further on the Greek etymology, "a twist . . . in the already twisted spatiotemporal fabric of an existing catastrophe" (71). Following either Morton's theoretical dazzle or Davies's icy clarity, we arrive in an Anthropocene that is no longer about deferring catastrophes but about enduring them, and building structures to address injustice as we do so.

Slavery as Culture, or Anthropos Self-Domesticated

The spiraling disasters that Davies and Morton extract from the geological record point toward political imperatives. One of Morton's recent books, *Humankind,* wears its utopianism in its subtitle: *Solidarity with Nonhuman People.*[9] Davies's goal of "ecological pluralism" (*Birth,* 208) and "keeping most of that remaining carbon in the ground" (207) are less abstract, but they also rely on political persuasion and other human cultural habits. Taking a long view, however, suggests that the structures that accompany civilization, including both agriculture and its cognate "culture," may not be up to the task. James Scott, in his recent book *Against the Grain,* reads current analysis of the archeological record to suggest that the earliest human states were not only physical disasters—hunter-gathers fed themselves better, did not suffer diseases of crowding, and enjoyed more leisure time than the first farmers—but the states that arose with the agricultural revolution could only grow through coercive means.[10] The key innovation, after the prehistoric technique of manipulating fire, was the "domus effect," which causes domesticated animals to develop different traits from

9. Timothy Morton, *Humankind: Solidarity with Nonhuman People* (London: Verso, 2017).

10. James Scott, *Against the Grain: A Deep History of the Earliest States* (New Haven, Conn.: Yale University Press, 2017).

their wild kin. Domestication would also be applied to humans as agricultural settlement expanded; "self-domestication" (83) would narrow human capacities and funnel their labors into agricultural monocultures. The practice of state formation did not invent human slavery, as far as the record shows, but early states did innovate, in Scott's phrase, "large-scale societies based systematically on coerced, captive human labor" (180). Or, more succinctly: "No slavery, no state" (156). To this self-domestication all modern states, democratic or not, are heirs. Building antislavery states turns out to mean building something new in human history.

Slavery states built the culture and eventually the industries that drove the Anthropocene. But anthropogenic climate change did not march in precise lockstep with the slaveholder. Developing (although not citing) the controversial "early Anthropocene" thesis of paleoclimatologist William Ruddiman, Scott makes the case for a "thin Anthropocene" traceable back to the earliest employment of fire by prehistoric hominids 400,000 years ago, substantially before the appearance of *homo sapiens* (19).[11] Scott's distinction between "thin" but accumulating climate disruptions decipherable in distant prehistory and a "thick" postindustrial Anthropocene datable through radioactive fallout after 1945 substitutes a continuous variety for the either-or debates of Golden Spike–ing. The thin–thick continuum opens up a plural way to conceive of the long history of anthropogenic climate change, which accompanies aggressive hominids throughout their history, and the global shock we now call the Anthropocene, which points specifically at the massive hockey-stick expansion of carbon emissions after industrialization. Anthropocene theorists should seek inside that

11. William Ruddiman, *Plows, Plagues, and Petroleum: How Humans Took Control of Climate* (Princeton, N.J.: Princeton University Press, 2005).

fissure a plurality of engagements and entanglements, not all of which need be only catastrophic.

Reading in and against the Anthropocene

What I'm proposing, modestly, amounts to a pluralized sense of *reading* under Anthropocene conditions. Here I take my cue from the valuable work of ecocritics Tobias Menely and Jesse Oak Taylor, for whose innovative collection *Anthropocene Reading* I am pleased to have written an ambivalent essay about the 1610 Anthropocene that represents my own farewell to Golden Spike-ing.[12] Giving to the stratigraphic signatures made legible by earth-systems scientists the deconstructive treatment familiar to post-Derridean generations of literary scholars, Menely and Taylor discover new possibilities: "the species that reads itself in the stone might yet be brought into a new degree of self-awareness *as* a species and, out of that recognition, weave new democracies and inclusive economies, conjoined to resilient ecologies" (21). Plural reading practices help imagine plural responses to ecological dynamism. The project Menely and Taylor imagine remains easier to conceive than to practice. Pluralizing confuses things we once thought we understood. None of the simple narratives I have introduced here—not Scott's deep history of slave states nor Usher's "untranslating" nor Neimanis's hydrophenomenology—can easily reach the daunting scale of those utopian tasks. Into plural futures we have little to offer besides *reading itself*, "an invariably polyglot, salvage practice" (13) that enables a partial and painful awareness of how it feels to live among physical changes unstoppably underway. "We read be-

12. Tobias Menely and Jesse Oak Taylor, *Anthropocene Reading: Literary History in Geologic Times* (State College: Penn State University Press, 2017). My essay in the volume is "Enter Anthropocene, circa 1610," 43–58.

cause we are terrified" (20), Menely and Taylor emphasize. Out of terror and interpretation we seek not a singular solution but a plural weave of possibilities. We seek justice and the alleviation of suffering, but not through assimilation of all thoughts and efforts into the capacious unity of the Borg. Differences can be painful and disruptive but plurality—in thinking and living, in ecosystems and politics—is worth striving for. Breaking up the Anthropocene will let different things come into view.

Some possible pluralities follow in these pages.

Six Human Postures

LIKE MANY OLDER MEN, Anthropos has problems with his posture. He tells himself the important thing is to stand up straight. From his exalted height, he towers over his rivals and can see past the horizon. But over the years it has become harder to maintain that upright position. His back aches. He feels the weight of history pressing down on the arches of his feet. Sometimes he thinks things would get better if he were just able to move his body into a different position.

The surge in eco-critical scholarship on the Anthropocene in recent years can be analogized to different postures for Old Man Anthropos. It's not easy at this point to place a substitute figure at the center of things; the ecological, cultural, and geophysical footprints the Old Man leaves in the sand aren't going away. But it's possible to conceive of postures beyond imperial erectness. Some of these positions even glimpse futures different from the bleak horizon Anthropos sees directly in front of his face.

Six postures provide a rough survey of the varied positions of ecotheory in the late twenty-teens. The most sinister vision sees Anthropos still Rising, looming over a future that's painfully like the present, only more so. An alternative posture imagines the Old Man Stooping to notice other forms of life and kinship relations. Political theorists imagine Anthropos straining to break his Chains and pluralize democracy for a more-than-human world. A different vector of materialism imagines a position of Amazement before the vorticular patterns and complex forms that may yet disrupt our headlong rush into disaster. A more thoroughgoing critique emerges from the ecofeminist critical Mirror, which emphasizes that the Old Man's masculinity always puts Nature and Woman on the same disadvantaged side. A more radical possibility takes the Old Man Unawares and finds new postures in Indigenous ideas and values. These half-dozen positions do not add up to a plan to resist Anthropos's dominion. But in their plurality and internal contradictions, they gesture toward things that He can't see.

Anthropos Rising

When he stands tall, it's bad news for the rest of us. But he makes an inviting target, with his eyes fixed forward. To many people, that massive body represents exploitation in its crudest form. Everything outside his body comprises material to be overseen, overcome, and processed. "Capitalism's governing conceit," writes Jason W. Moore, "is that it may do with Nature as it pleases, that Nature is external and may be coded, quantified, and rationalized to serve economic growth, social development, or some other higher goal."[1] The transforming task that upright Anthropos accomplishes is a particular vision: he sees the world as Nature, resource, or what Heidegger famously calls "standing-reserve."[2] As Moore transforms Marxist economic

1. Jason W. Moore, *Capitalism in the Web of Life: Ecology and the Accumulation of Capital* (London: Verso, 2015), 2.

2. Martin Heidegger, "The Question Concerning Technology," in *Basic Writings, ed.* David Farrell Krell, 307–42 (New York: Harper Collins, 1993)

history into what he calls "*world-ecology*" (3, italics in original), he provides a clear narrative in which Anthropos devises the rapacious system called Capitalism and thereby fouls the world. In Moore's reading, capitalist rupture begins not with the Industrial Revolution but several centuries earlier, with the global expansion of European economies and populations following the so-called Columbian Exchange that kicked off the catastrophes of ecological globalization.[3] Expanding beyond its geographic boundaries enabled European cultures to operate through what Moore terms the strategy of "Cheap Nature," which uses external resources, including slave labor as well as nonhuman abundance, to fuel geopolitical expansion. But the commanding vision Moore would have us rename Capitalocene does not confine itself to the merely economic. Instead, he argues that "at the heart of modernity's co-production is the incessant reworking of the boundaries between the human and the extra-human" (17). Through repeated processes of "frontier-making" (66)—Moore notes the repetition of strategies in European expansion in Ireland, the Canaries, the Caribbean, the Americas, and the American West—elements of Nature are made Cheap and then exploited. As Moore puts it in a book coauthored with economist Raj Patel, "Cheap is a strategy, a practice, a violence that mobilizes all kinds of work—human and animal, biological and geological—with as little compensation as possible."[4] Cheap burns out from Anthropos's keen eyes and reduces the world.

3. Alfred W. Crosby, *The Columbian Exchange: Biological and Cultural Consequences of 1492* (Westport, Conn.: Greenwood Publishing, 1972). For my speculations on why "ecological globalization" may be a preferred term, see *Shipwreck Modernity: Ecologies of Globalization, 1550–1719* (Minneapolis: University of Minnesota Press, 2015), xxvi–xxxii.

4. Raj Patel and Jason W. Moore, *A History of the World in Seven Cheap Things: A Guide to Capitalism, Nature, and the Future of the Planet* (Berkeley: University of California Press, 2017), 22.

Moore's vision of Anthropos-as-Capital emphasizes the thoroughness of a five-hundred-year exploitative run that may, in the early twenty-first century, have run out of new frontiers. His analysis links resource exploitation with slavery—he and Patel posit that modern slavery, as distinct from classical slavery, was born in Madeira with the first sugar plantations (*History*, 30)–as well as racial and gender discrimination. But as tightly interwoven and rapaciously efficient as the system has been over the past half-millennium, the current environmental crisis holds the seeds of its own destruction. In dialogue with other environmental theorists including Donna Haraway, Moore argues that "capitalism's either/or organization of reality" into "Nature/Society" may be giving way to a recognition of "human organization as utterly, completely, and variably porous within the web of life."[5] Patel and Moore find hints of alternatives to European ecological exploitation in the dream of an anonymous Chichimec woman hanged by Spanish colonizers in Mexico in 1599 (*History*, 44–46). Moore locates the seeds of revolutionary change in "alternative valuations of food, climate, nature, and everything else" (*Anthropocene or Capitalocene*, 11). The conclusion of *A History of the World*, his collaboration with Patel, sings out with alliterative optimism: recognition leads to reparation, redistribution, reimagination, and finally a thoroughgoing "recreation" of work, leisure, and the human relationship with the nonhuman environment in a "reparation ecology" (202–12). In such a vision, Old Man Anthropos may still be standing, but his view no longer threatens.

The relatively sanguine conclusion at which Patel and Moore's eco-history arrives glides past the more violent futures that many Marxist intellectuals see emerging out of Anthropocene capital-

5. Jason W. Moore, ed., *Anthropocene or Capitalocene? Nature, History, and the Crisis of Capitalism* (Oakland, Calif.: Kairos/PM Press, 2016), 5.

ism. A helpful counterpoint to Moore's world ecology appears in McKenzie Wark's *Molecular Red: Theory for the Anthropocene.*[6] Taking his inspiration from the Soviet radicals Alexander Bogdanov and Andrey Platonov, whose work he connects to American writers Donna Haraway and Kim Stanley Robinson, Wark asks critical theory to "*change labor,* by merging art and work; to *change everyday life,* by developing the collaborative life within the city and changing gender roles and norms; and to *change affect,* to create new structures of feeling" (35, italics in original). Wark reads Bogdanov's *Essays in Tektology* (1912–17) as "science fiction in reverse": "Bogdanov wrote for the Martians, and the Martians are us" (59). In returning his radical thinking to its roots in Soviet communism and social experiment, Wark champions what he calls "molecular" thinking: "For it is the molecular scale which corresponds best to the labor point of view" (219). Extending the point, he argues for a turn away from theoretical niceties toward material realities: "now perhaps what we need is a pungent dose of vulgarity" (220). Unlike Patel and Moore, who in essence ask our present civilization to see reason and stand down from its domineering perch, Wark offers the disorienting dream of "meta-utopia . . . not so much an imaginary solution to real problems as a real problematizing of how to navigate the differences between the imaginal that corresponds to each particular labor points of view" (225). Wark's opening-into-difference rejects the capitalist present for an almost-unimaginable future. "We all know this civilization can't last," he quips. "Let's make another" (225). The shape of that "other" remains opaque—but it's certainly not the shapes Old Man Anthropos sees in the horizon he thinks of as his own.

Somewhere between Moore and Patel's rational eco-Marxist

6. McKenzie Wark, *Molecular Red: Theory for the Anthropocene* (London: Verso, 2015).

analysis and Wark's radical futurity Anthropos stands tall, weathering the storm, as least so far. It seems impossible that this posture can last.

Anthropos Stooping

What might Anthropos see if he looks down below his too-large feet? In today's "disturbance-based ecologies," the answer, if he's lucky, will be mushrooms.[7]

Anna Tsing's eloquent and carefully antistructured book *The Mushroom at the End of the World* explores the late twentieth-century flourishing of the international trade in Matusutake mushrooms, which thrive in disturbed forests such as those in heavily logged rural Oregon and are beloved by connoisseurs in Japan. The mushrooms and the temporary cultures of harvest, sale, and transport that arise around them help her to "open our imagination" (19) to new transcultural and transpecies exchanges in the postmodern world. In arguing that "precarity . . . being vulnerable to others" (20) is "the condition of our time" (20), she celebrates "salvage accumulation" (63). Alongside her mushroom hunters, whose numbers include hill peoples displaced from Southeast Asia, loggers displaced by the collapse of an industry, hippies, and assorted other vagabonds, she discovers in the ruins of natural and industrial ecologies the seeds of new and fecund life. Dancing between large-scale and local histories with an anthropologist's deft deferral to primary actors, she ends her book with a chapter called "Anti-ending" and admits that her argument "does not properly conclude" (378). Having opened by slyly insinuating that "unlike most scholarly books" she will offer "a riot of short chapters" (viii) rather than denser

7. Anna Tsing, *The Mushroom at the End of the World: On the Possibility of Life in Capitalist Ruins* (Princeton, N.J.: Princeton University Press, 2015).

fare, she ends in a woodland ecology that is both material and metaphorical, circulating both mushrooms and ideas.

An important collaborator with and influence on Tsing's scholarship, her Santa Cruz colleague Donna Haraway provides perhaps the clearest feminist redirection of the macho pieties of Anthropocene thinking. Haraway, in her latest volume, *Staying with the Trouble,* offers her own Chthulucene as alternative and insists on the slogan "Make Kin, not Babies!"[8] With the acerbic wit that has defined her writing since "A Cyborg Manifesto" first appeared in 1984, Haraway rejects the doom language of Anthropocene theorizing in favor of an entangled, multispecies view. She also rejects "posthuman" as label and theoretical position, wanting to be down in the dirt rather than up in airy metaphors. In a recent conversation with Cary Wolfe, she proposes two rival etymological chains for the term human. "Too many" understandings of the terms, she argues, "go to *homo*—which is the 'bad' direction," but it's also possible to value the "'human' that goes to *humus,* which is the 'good' direction."[9] In the contrast between "soil" and "the phallic 'man'" (261) she generates another slogan: "Not Posthumanist But Compost" (262), with the final word ambiguously hovering between being a noun—to be a composted thing—and an imperative verb—to compost all things together in order to make something new. In both meanings, Haraway's compost requires intimate and self-changing mixtures. We don't end up the same being that we were when we started.

When he's standing tall, Anthropos can't see the mushrooms and composting reaches him only as a faint acrid smell. In order to see these things, he must stop, stoop, and pay attention.

8. Donna Haraway, *Staying with the Trouble: Making Kin in the Chthulucene* (Durham, N.C.: Duke University Press, 2016), 5–6.

9. Donna Haraway and Cary Wolfe, "Companions in Conversation," in *Manifestly Haraway* (Minneapolis: University of Minnesota Press, 2016), 201.

Anthropos in Chains

Perhaps the most famous utopian image of the political discourses of the ecotheoretical turn has been Bruno Latour's "Parliament of Things," which redefines "the continuity of the collective" so that humans and nonhumans alike gain representation in a messy democratic conclave: "Let one of the representatives talk, for instance about the ozone hole, another represent the Monsanto chemical industry, a third the workers of the same chemical industry, another the voters of New Hampshire, a fifth the meteorology of the polar regions . . ."[10] The fantasy of a world that speaks for itself, in all its messy pluralities, would become the spine of Latour's Actor-Network Theory and an animating force in his ongoing work on environmental politics in the present.[11] Latour now claims to prefer the three-part term "New Climatic Regime" over the omnipresent Anthropocene, but he also implies that responding to the new-ness (i.e., the 'cene-ness) of our present will require all the resources of the humanities, social sciences, and other fields. To craft a new metalanguage, Latour rejects abstract formulations such as "modernity," the Anthropocene, and even "globalization," in order to find in the material conditions of "Gaia" a structure that accommodates complexity and what Latour calls (adapting the language of Isabelle Stengers) "sensitivity."[12] In rejecting "the two

10. Bruno Latour, *We Have Never Been Modern,* trans. Catherine Porter (Cambridge, Mass.: Harvard University Press, 1993). French original, 1991.

11. On Actor-Network Theory, see Bruno Latour, *Reassembling the Social: An Introduction to Actor-Network Theory* (Oxford: Oxford University Press, 2007). For Latour on the Anthropocene, see *Facing Gaia: Eight Lectures on the New Climatic Regime,* trans. Catherine Porter (New York: Polity Press, 2017).

12. See *Facing Gaia,* 136–45. See also Isabelle Stengers, *In Catastrophic Times: Resisting the Coming Barbarism,* trans. Andrew Goffey (Ann Arbor, Mich.: Open Humanities Press, 2015).

great unifying principles—Nature and the Human—[as] more and more implausible" (*Facing Gaia,* 142), Latour turns into a deeper complexity that his Parliament of Things had initially proposed. "Once the Globe has been destroyed," he writes, "it has space and time enough so that history can start up again" (*Facing Gaia,* 145). Even more recently Latour names his preferred new impulse "Terrestrial."[13] The shapes of the new earth history may include not just the babblings of an impossibly polylingual Parliament but also sensitivities to wonder that occupy the carved-out space of the literary sublime.

As Latour's speculations have grown increasingly baroque, other thinkers have taken up the task of rationalizing his structures. The most persuasive political account of how democracy might adapt to the posthuman and post-Nature circumstances of the Anthropocene comes in American legal scholar Jedediah Purdy's *After Nature: A Politics of the Anthropocene.*[14] Seeking what he calls a "*Walden* for the Anthropocene" (147–52), Purdy argues that our era of climate disruptions "radicalizes eco-awareness into a fully democratic politics of nature" (206). This effort encounters a series of problems: skepticism, particularly in relation to technical expertise (268–69), utopianism in regard to democratic openness (269–70), and the frustrations inherent in "bridging . . . the present state of things with the idea of a democratic Anthropocene" (270). Like Latour, Purdy finds inspiration in environmental art and literary culture; he hopes to "make the imaginative literature of an Anthropocene democracy serve as a productive fiction" (270). Bringing the theoretical speculations of Latour and the posthuman materialism of Jane Bennett in touch with political realities, including the oversized role of money

13. Bruno Latour, *Down to Earth: Politics in the New Climatic Regime* (Cambridge, UK: Polity Press, 2018), 40.

14. Jedediah Purdy, *After Nature: A Politics of the Anthropocene* (Cambridge, Mass.: Harvard University Press, 2015).

in twenty-first century American politics (271), Purdy seeks to reconcile the egalitarian aims of humanism with a posthuman expansion of the franchise in order to create "a democracy open to the strange intuitions of post-humanism: intuitions of ethical affinity with other species, of the moral importance of landscapes and climates, of the permeable line between humans and the rest of the living world" (282). This "democracy open to post-human encounters" (288) asks that the Anthropocene disrupt our politics as well as our environments.

From the gloomy perspective of the political landscapes of Europe and North America in 2019, it's hard to see how we get to Purdy's Anthropocene democracy or to the more dizzying channels of Latour's Gaia-inflected polities. Anthropos today, to borrow Rousseau's famous Enlightenment formulation, appears everywhere in chains. But Latour's insight about the hollowness of "the modern," which he conceived after the fall of the Berlin Wall in 1989 (*We Have Never Been*, 8–10), emphasizes that our current transition into the Anthropocene is less a choice that a rupture with a fantasized version of our own past. "We scarcely have much choice," he observes, "It is up to us to change our ways of changing" (145). To free Anthropos, we must lean into dynamism. It may be frightening, and it certainly will be disorienting. But futures beckon, beyond the constricted realms of back-facing nostalgia.

Anthropos Amazed

No ecotheorist has generated more heat and wonder than the voluminous Timothy Morton, author of six major books of ecotheory in the past decade.[15] Morton's churning productivity and

15. Morton's eco-books include *Ecology without Nature* (Cambridge, Mass.: Harvard University Press, 2007); *The Ecological Thought* (Cambridge, Mass.: Harvard University Press, 2010);

occasional repetition of his central theoretical moves has earned him some hostile reviews, but his insight into how ecological thinking implicates humanity in the more-than-human environment remains influential. While he claims to reject the poetic sublime that was the poetic mode of Percy Shelley, the poet on whom Morton cut his academic teeth, to some extent his cornucopia of neologisms—hyperobjects, dark ecology, the strange stranger, agrologistics, the Severing, et cetera—might productively be imagined as redeploying the Romantic sublime in nonanthropocentric fashion. A recent ringing pronouncement claims that "Anthropocene is one of the first truly anti-anthropocentric concepts because via thinking the Anthropocene, we get to see the concept of 'species' as it really is—species as a subscendent hyperobject, brittle and inconsistent."[16] In his most recent book, he has argued that the preferred term for general use should be "mass extinction," which gets to the point more forcefully than "global warming" or "climate change."[17] The restless energy of such flashy coinages signals Morton's desire, which is not inconsistent with the revolutionary poetics of his one-time master Shelley, to "keeping the future open" (*Humankind,* 153) even as doom gets pronounced at every turn. In the climate change–ruptured present in which "the basic mode of ecological awareness is anxiety," art becomes a formula for "grief-work."[18] Morton's tragic intensity testifies to his post-Romantic restlessness and ambition. He wants, since we cannot make Anthropos sit still, to make him sing.

From the point of view of the stooping mushroom pickers and kin makers Tsing and Haraway, Morton's neo-sublime

---

Hyperobjects: Philosophy and Ecology after the End of the World (Minneapolis: University of Minnesota Press, 2013); *Dark Ecology* (2016); *Humankind: Solidarity with Nonhuman People* (London: Verso, 2017); and *Being Ecological* (Cambridge, Mass.: MIT Press, 2018).

16. Morton, *Humankind,* 113.

17. Morton, *Being Ecological,* 5.

18. Morton, *Dark Ecology,* 130; Morton, *Hyperobjects,* 196.

ecothinking appears excessive, or at least overly masculine in its eagerness to be always visible. Within premodern literary ecostudies, a compelling alternative discourse of vorticular entanglement has emerged through the collaboration of medievalist Jeffrey Jerome Cohen and early modernist Lowell Duckert. In editing together and separately four collections—a special issue of the journal *Postmedieval* and a trilogy of ecotheoretical books—Cohen and Duckert have historicized and pluralized Anthropos's wonderment.[19] When Cohen claims that "catastrophe is entanglement," he rewires the human relationship with forces in our nonhuman environment that do us hurt.[20] Both ecocritics turn toward difficulty with eager and imaginative desire. Their shared obsession with vortexes structures their investigation of matter and metaphor. "The Shape of the Elements is a vortex" (20) they emphasize in *Elemental Ecocriticism*. In the moving and also briefly still center of the storm, they conspire forward into the Anthropocene.

The project these two critics have jointly assembled emerges in its greatest power when it gives itself away. Inverting Morton's sometimes manic efforts to roll all aspects of "the ecological thought" into a single vibrant package, Cohen and Duckert invite other creators, human and nonhuman, into community. In the

19. The collaboration started with the "Ecomaterialism" special issue of *Postmedieval* 4, no. 1 (Spring 2013) and then extended through three collections: *Prismatic Ecology: Ecotheory beyond Green* (2013), ed. Jeffrey Jerome Cohen; *Elemental Ecocriticism: Thinking with Earth, Air, Water, and Fire,* ed. Jeffrey Jerome Cohen and Lowell Duckert (2015); and *Veer Ecology: A Companion for Environmental Thinking,* ed. Cohen and Duckert (2017), all published by the University of Minnesota Press. These four volumes include the work of roughly fifty authors in multiple fields. (Full disclosure: I have an essay in each of the four collections.)

20. Jeffrey Jerome Cohen, *Stone: An Ecology of the Inhuman* (Minneapolis: University of Minnesota Press, 2015), 65.

introduction to *Prismatic Ecology,* Cohen emphasizes that "the Mississippi [river] is an earth artist, but its projects take so long to execute that humans have a difficult time discerning their genius" (xix). The brown sinuous flows of the river underlie the pluralization of green thinking that emerges in *Prismatic Ecology* and continues through the collaborations with Duckert.[21] In the four classical elements, these two authors discover ways to reinvent catastrophic entanglement: "Catastrophe is a kind of 'forward-thinking' in search of more capacious futures, a drama of unidentifiable genre, a tragicomedy that picks up where the 'comedy of survival' left off."[22] Bringing together Love and Strife, the two principles that for the classical philosopher Empedocles control relations among the elements, these critics recast opposition and struggle as collaboration and (in a term Duckert adapts from Latour) "composition."[23] "Empedocles wrote of pervasive disordering force," they recall, "only to move to an emphasis on that which binds."[24] In Cohen and Duckert's example, it's the moving that matters, the constant shifting of perspectives and ceding centrality to other voices.

How does this collaborative exfoliation speak to Anthropos's changing postures? While Morton, in my view, maintains a neo-Romantic urgency that perhaps he draws from Shelley's revolutionary verse, Cohen and Duckert produce an Anthropos who neither rages nor roars. Their project, as they write in *Veer*

21. For Cohen's critique of "Green Criticism," see *Prismatic,* xix–xxiii.

22. *Elemental,* 18. This passage plays with two landmark ecocritical projects: "forward thinking" is the title of Morton's third chapter in *The Ecological Thought,* and the "comedy of survival" refers to Joseph Meeker's early study, *The Comedy of Survival: Literary Ecology and a Play Ethic,* 3rd ed. (Tucson: University of Arizona Press, 1997).

23. See *For All Waters, passim.* Duckert draws on Latour's "An Attempt at a 'Compositionist Manifesto,'" *New Literary History* 41 (2010): 471–90.

24. Jeffrey Jerome Cohen and Lowell Duckert, "Howl: Editors' Introduction" *Postmedieval* 4, no. 1 (Spring 2013): 1–5 at 5.

Ecology, is to "render snug habitations strange, opening them to a world agentic and wide" (11). Like modern ergonomic desk chairs that encourage movement and position-shifting throughout the workday, their Anthropos changes postures. At one point, he may have enjoyed the promise of the rainbow. But soon fire drives him out toward roiling waters, and those directions also veer into newness in their most recent collaboration. To make Anthropos plural, he must remain just a bit on edge, not always comfortable, never still.

Anthropos in the Mirror

When Anthropos looks in the mirror, he sees, if he's paying attention, a woman's face. Centuries of patriarchal conceptions of "Man" and "Nature" have labored to obscure the place of gender in environmental thinking, but attending to ecofeminist voices can help correct mono-masculinity. In opposition to every macho mountain man like John Muir, a feminist ocean poet like Rachel Carson raises her face into view. Like Tiresias, Old Man Anthropos spans both genders.

Ecofeminism sometimes articulates itself as critique of patriarchal assumptions. In the phrase of New Media scholar Joanna Zylinska, the core mandate for human thinking in the Anthropocene is to "tell better stories."[25] Stacy Alaimo concurs with the queer-inflected admonition that "the Anthropocene is no time to set things straight."[26] Ecofeminism requires Anthropos to change postures, to move and respond to what Alaimo elsewhere calls the "profound shift in subjectivity" occasioned by

25. Joanna Zylinska, *Minimal Ethics for the Anthropocene* (Ann Arbor, Mich.: Open Humanities Press, 2014), 46.

26. Stacy Alaimo, *Exposed: Environmental Politics and Pleasures in Posthuman Times* (Minneapolis: University of Minnesota Press, 2016), 1.

recognizing the permeability of bodies and environments.[27] Living amid the cultural toxicity of patriarchy and the material toxicity of late capitalism invites twenty-first-century human subjects to embrace what Richard Grusin, editor of the book *Anthropocene Feminism,* calls an "ethic of disruption."[28] In a utopian spirit, Grusin imagines his multivoiced collection as an "assembling of small-scale systems or the claiming of responsibility for all human and nonhuman actants toward a goal of mutual thriving" (xi). For Anthropos and for ecological thinking in the Anthropocene, such a variety of voices generates important pressures and desires. No one wants to return to monolithic "Man" in conflict with equally colossal "Nature." Anthropocene feminism opens doors to the multitude of perspectives that can replace these monuments. No strands within Anthropocene thinking turn more directly into plurality than feminist discourses.

At the core of theoretical ecofeminism, and also at the core of many discourses of twenty-first-century environmentalist thinking, sits the ambiguous figure of the human. Rosi Braidotti stakes out the clearest critical position when she writes, "feminism is *not* a humanism" (*Anthropocene Feminism,* 21). Moving into radical "species egalitarianism" (32), Braidotti imagines the benefits that flow from rejecting old-fashioned ideas of human centrality: "We may yet overcome anthropocentrism by becoming anthropomorphic bodies without organs that are still finding out what they are capable of becoming" (35). Adapting Delueze and Guattari as well as Jane Bennett's call for "strategic anthropomorphism," Braidotti joins figures such as Cary Wolfe to push the post-

27. Stacy Alaimo, *Bodily Natures: Science, Environment, and the Material Self* (Bloomington: Indiana University Press, 2010), 20.

28. Richard Grusin, ed., *Anthropocene Feminism* (Minneapolis: University of Minnesota Press, 2017), xi.

humanist edge of ecotheory.[29] By contrast, ethicist Zylinska emphasizes the "strategic role of the concept of the human in any kind of ethical project worth its salt" (61). The suggestive concept of "ecotone war" currently being developed by Joshua Clover and Juliana Spahr, implies that environmental conflict—their ur-example is the material struggle between land and sea—models political conflicts yet to come.[30] They see resistance to gender inequality and capitalist exploitation as "conjoined struggles" (166), thus suggesting that the gender identity of Old Man Anthropos must be exploded, rather than contained or even bifurcated. On the subject of the humanity of Man, it seems, ecofeminist thinking remains unsettled.

Another shifting context that appears productive for ecofeminists is the shift from the purely terrestrial to mixed land-and-sea perspectives. Clover and Spahr examine strikes in California port cities during the brief "Occupy" heyday in 2011 as revealing the land–sea rift. Alaimo argues for "dwelling in the dissolve" in which aqueous environments threaten fictions of bodily solidarity.[31] The "loss of sovereignty" of dissolution amounts, in her terms, to "an invitation to intersubjectivity or trans-subjectivity and even . . . to a posthumanist or counter-humanist sense of self as opening out unto the larger material world and being penetrated by all sorts of substances and material agencies that may or may not be captured" (*Exposed,* 4). Relatedly, Astrida Neimanis treats the human body as a form

29. Gilles Deleuze and Félix Guattari, *A Thousand Plateaus: Capitalism and Schizophrenia,* trans. Brian Massumi, 2nd ed. (Minneapolis: University of Minnesota Press, 1987); Jane Bennett, *Vibrant Matter: A Material Ecology of Things* (Durham, N.C.: Duke University Press, 2010).

30. Joshua Clover and Juliana Spahr, "Gender Abolition and Ecotone War," in *Anthropocene Feminism,* ed. Grusin, 147–67.

31. Alaimo, *Exposed,* 168. This chapter, "Your Shell on Acid," also appears in *Anthropocene Feminism.*

of membrane: "This membrane is not a discursive barrier, but an interval of passage: solid enough to differentiate, but permeable enough to facilitate exchange."[32] In broader cultural and philosophical context, Karin Amimoto Ingersoll has presented a "seascape epistemology," which is "not a knowledge of the sea . . . [but] a knowledge about the ocean and the wind as an interconnected system that allows for successful navigation."[33] Her focus on intentional movement—navigation—through saltwater spaces connects her surfing perspective to the global navigational techniques of Polynesian mariners and European chart makers. For all these writers, turning away from green terrestrial pastoralism toward blue oceanic dynamism enables a new perspective on the Anthropocene. The turn from the solid ground of modernity's petro-expansion to the global currents of oceanic immersion may represent a productive conceptual seascape in which to imagine Anthropocene futures.[34]

Anthropos Unaware

This archive of postures and methods is not as plural as I wish it could be. The challenge of pluralizing is that it's never quite enough: further changes always beckon. I can imagine pos-

32. Neimanis, *Bodies of Water,* 98. Neimanis develops her idea of membrane in dialogue with Karen Barad's "agential separability" in *Meeting the Universe Halfway: Quantum Physics and the Entanglement of Matter and Meaning* (Durham, N.C.: Duke University Press, 2007) and Luce Irigaray's figurations of water in *Marine Lover (of Friedrich Nietzsche)* trans. Gillian C. Gill (New York: Columbia University Press, 1991). For my own speculations about seepage across borders, see "Seep," in *Veer Ecology, ed. Cohen and Duckert,* 282–96.

33. Karin Amimoto Ingersoll, *Waves of Knowing: A Seascape Epistemology* (Durham, N.C.: Duke University Press 2016), 6.

34. On Ocean-as-Anthropocene, see my entry in *Anthropocene Unseen: A Lexicon,* ed. Cymene Howe and Anand Pandian, (Brooklyn, N.Y.: Punctum Books, forthcoming).

tures for urban ecologies, queer ecologies, terraqueous strains, and cyborg assemblages. Even more enticing are the postures that I cannot imagine, those that take the Old Man Unawares. Probably the biggest gap in my own expertise lies in non-Western and Indigenous environmental thinking. These ideas present important challenges to the singularity of the Old Man's current postures. Professor and activist Kyle Whyte argues that "Anthropogenic (human-caused) climate change is an intensification of environmental changes imposed on Indigenous people through colonialism."[35] Whyte claims that Indigenous thinking and scholarship provides an essential counterdiscourse to pluralize our ideas about humans and their nonhuman surroundings. These ideas cannot fail to expand our postures and visions.

In some cases, mainstream scholars have begun to bring the Unaware into familiar discourses. One recent scholarly example appears in philosopher Jonathan Lear's 2006 book, *Radical Hope,* which unfolds the story of Plenty Coups, the nineteenth-century Native American Crow leader who guided his people to accept the end of their traditional way of life. Plenty Coups's dilemma—"How ought we to live with this possibility of collapse?" (9)—resonates with the dire pronouncements of environmental doomsayers in the Anthropocene.[36] In Lear's reading, Plenty Coups shows that it's possible to reframe breakdown as futurity. "We must do what we can," Lear ventriloquizes the Crow leader, "to open our imagination up to a radically different set of future possibilities" (93). Plenty Coups's vision of the Crow people enduring without mobility,

35. Kyle Whyte, "Indigenous Climate Change Studies: Indigenizing Futures, Decolonizing the Anthropocene," *English Language Notes* 55, no. 1–2 (Fall 2017): 153–62.

36. Jonathan Lear, *Radical Hope: Ethics in the Face of Cultural Devastation* (Cambridge, Mass.: Harvard University Press, 2006).

wealth, or war parallels our diminished prospects in the age of climate change.

Lear's analysis of the dilemma of Plenty Coups sits uneasily alongside Whyte's present-day environmental activism in the service of indigenous climate justice. To respond to Indigenous thinking may require a deeper reckoning with activism as well as an expanded critical imagination. As I continue to learn about Indigenous responses to the Anthropocene, I anticipate finding more, and more challenging, postures into which Old Man Anthropos must bend his body.

What happens if we put all the postures together?

There's no simple pattern Anthropos can master. *Stooping* after *Rising* captures a pleasing symmetry, but the political *Chains* in which Anthropos finds himself inhibit further movement. He can't find his rhythm. At times *Amazed* by the wideness of his world, he finds in the *Mirror* a gendered plurality that even oceanic depths and remote deserts have not revealed to him. To a large extent he remains *Unaware*. He wants to put all the steps together, to build new alliances through these plural postures and to expand his repertoire to new wanderings and fresh pastures. History and materiality together produce so many constraints. Can he do it?

Dance, Old Man! Dance!

Anachronism as Method

THE BAD ONE is comforting, false, deceptive, and more often than not deeply desired. It says things some people find easy to believe. In the United States after 2016, bad anachronism wears red baseball caps and threatens people who it doesn't recognize.

The good one is disorienting, challenging, true, difficult, and hard to wrap your imagination around. It reminds us of things we might sometimes wish to ignore. This attitude toward history and social change may seem on the run just now, but it'll be back. The arc of history bends into complexity.

In volume 1 of *The Beast and the Sovereign,* Derrida makes a facile-seeming but quite useful distinction between "bad" and "good" anachronism.[1] The deconstructionist ambiguity at which the philosopher eventually arrives seems both predictable and convincing: "every reading is [. . .] anachronistic." But juxtaposing historicist rigor and nostalgic blindness isn't simple or straightforward. Subtle slips lurk between the good and the

1. Peter Adkins brought Derrida's passage to my attention in a perceptive review of my book *Shipwreck Modernity*: https://glasgowreviewofbooks.com/2016/07/20/anthropocene-flotsam-steve-mentzs-shipwreck-modernity-ecologies-of-globalization-1550-1719/. Accessed July 19, 2018.

bad. Whenever a scholar of premodern literature like me invokes the Anthropocene, a term popularized after the chemist Paul Crutzen coined it in 2000, we court anachronism—though the term also invokes a host of previous "Ages of Man," from Renaissance humanism to the rise of sculptural naturalism in classical Greece.

I seek a historically aware environmental humanities that can distinguish between good and bad anachronism. I want anachronism to help me think about humans, humanism, and the humanities during this period in the twenty-first century when the political forces of brute nostalgia and bad anachronism—what Timothy Snyder calls the "politics of eternity"[2]—wax stronger in the United States and Europe. Scholars and citizens need good historical examples, and we also require potent ways to fight back against the smothering nostalgia of eternity. We need good, factual historicism, and also responsible imaginative anachronism. The fortress of historical accuracy cannot protect humanist or posthumanist ideals. Historicist ecocritics especially need to anachronize positively, to use good and messy anachronisms to challenge bad and comforting nostalgia.

All the most valuable concepts in the environmental humanities flirt with anachronism, because they speak to at least two times at once. Anthropocene thinking uses a new geophysical designation to reframe received narratives about the relationship between humans and the nonhuman world. Such thinking dances up close to anachronism, stares deeply into its variably chronic eyes. What would it mean for anachronism to say yes to history? Or for historical scholarship to assent to anachrony?

2. Timothy Snyder, *On Tyranny: Twenty Lessons from the Twentieth Century* (New York: Penguin, 2017).

What polychromic chimeras are even now slouching toward turbid oceans to be reborn?

In mainstream literary scholarship today, as successive generations of graduate students have learned to their peril, anachronism serves as a cudgel with which to beat one's rivals. When I was a grad student in early modern literature in the '90s, anachronism seemed an unforgiveable error, the thing that would pull back the curtain and reveal me as scholarly imposter. I value the old and obscure as much as the next premodern scholar, and I would never forego the mysteries and pleasures of the archive, but we need a better language to define productive anachronism. The bugbear of "presentism," which in different contexts can be either accolade or accusation, doesn't sufficiently describe the multichronic perspective that literary scholarship brings to bear.

"The bug which you would fright me with, I seek," says distraught Hermione on trial in *The Winter's Tale*.[3] What if anachronism isn't a threatening slip into the nonscholarly abyss but an active challenge? What does anachronism want? What would anachronism do? (WWAD?) Humanities scholars need ways to promote good anachronism and resist its bad twin. Should we celebrate polytemporal lures or semi-legible palimpsests? Pay court to diachronic parallels? Resist, like Spenserian knights, the twin temptresses of past and present?

The dilemma is particularly acute for historicist and literary ecocritics, since our subfield can't help but keep a Cyclopean single eye fixed on looming ecological catastrophe in the twenty-first century. Those of us who seek to pluralize the Anthropocene want to disperse environmental thinking beyond the postindustrial or post-1945 timeframes, but the hot oceans and parched deserts of the twenty-first century force themselves into view.

3. William Shakespeare, *The Winter's Tale,* ed. John Pitcher (London: Bloomsbury/Arden3, 2010), 3.2.90.

Living in our time, we can't help but be anachronistic—that's Derrida's unarguable point—so perhaps it's time for us to come out and admit it?

I'm not ready to devise a comprehensive solution for anachronism policing, except perhaps to suggest that we do less of it, or at least employ less gleeful aggression when exposing its errors. A touch of empathy might benefit all reviewers, anonymous or otherwise. The Gordian time-knot of past and present resists unpicking, and the Alexandrine solution seems reckless. But perhaps a few principles are in order? Some goals to make our anachronizing more productive?

1. Anachronize now!

It is always the right time. The past is fluid, vast, and significantly unknown, but its presence leeches into our present each instant. The urgent *now* of lived experience undergirds all understanding, including that of historical scholarship; the sensation of living *now* anchors and motivates all our thinking, as it motivated the thinking of the historical actors who preceded us. The tension between anachronism's multiplicity and *now*'s single precision helps remind us that, in the good anachronism, nothing stays still. *Now* moves and shifts, as different temporalities adjust themselves in dialogue with each other and with changing audiences. Anachronizing now commits us to discourses of change and worlds that do not remain the same. The constant and inconstant principle of change enables us to resist the imaginary stasis of the past's eternal glories.

2. Anachronism's opposite is not historicism but "timelessness."

Good anachronism does not refuse historicism's rigor but instead rebuts the fantastical and destructive dream of timeless ideals. The problem to be defeated is less the brute force of historical distance than the illusion that certain ideologies are "always" valid. Nostalgia can be pernicious. It's the bad anachronism of falsifying nostalgia that encourages white men to believe their right to political dominion is sacrosanct. That same nostalgia insists that certain canonical figures—say Shakespeare or Virgil—represent "timeless human nature" about which nothing can or should change. A richly anachronized historicism combats timelessness with variety. Canonical literature unveils plurality, not eternity.

3. Anachronism favors utopia.

The reason to engage the past is to build a better future. Against the "again" in the backward-looking imperative to "Make American Great *Again*," anachronism opens an untrod pathway toward utopia. One way of voicing that hopeful social ideal has recently turned a half-millennium old, with the five-hundredth birthday of Thomas More's *Utopia* in December 2016. Via a famously bifurcated etymology, utopia is plural at it source; the word means both "no place" (*ut topos*) and "beautiful place" (*eau topos*). Good anachronism creates plurality; bad anachronism locks down eternal sameness. The road to utopia turns often.

"Now, Now, Very Now!"

And then I thought: each thing follows each other thing at the same time: precisely, precisely *now*.

—JORGE LUIS BORGES, "The Garden of the Forking Paths"

FIXING THE FEEL OF TIME in the Anthropocene may only be possible with help from the poets. Here, recklessly mixed, are three of my favorites: William Shakespeare, Emily Dickinson, and Jorge Luis Borges. They give access to the riotous pluralities that make up the present instant.

What time is the Anthropocene? Iago's time: "Now, now, very now!" (1.1.87).[1]

This *now* is out of joint. That's what Anthropocene means: not that humans control the world, but that we have tangled our contacts with personal and geological time. The time of *now* is not a displaced bone that can be put right, not an inexplicable swerve, not a passing phase. Ideas and objects dissolve under our fingers. Fantasies in which culture exists apart from nature disintegrate. However we look, backward or forward into catastrophe, the leering visage of Old Man Anthropos looms. He corrupts everything, including the past. But he isn't—which is to say, we, or at least some of us, aren't—monolithic. Global climactic disruptions accent human frailty. Our dissolved and dissolving environment validates the posthuman critique that kneecaps Man's dreams of power. But the

1. William Shakespeare, *Othello,* ed. E. A. J. Honingmann (London: Bloomsbury/Arden 3, 2003). Quotations in the text by act, scene, and line numbers.

human does not crave mastery only. Humanism and the humanities also speak for a dream of justice to which it is worth clinging, even if we can no long credit the control fantasies of any Age of Man. Living in Anthropocene dissolution requires inventing new ways to experience time. Why not swim in time or fly through it, since living at a plodder's pace seems so constricted?

Anthropocene names a grinding reality and perhaps an impossible task, but its temporality swirls in both directions, as Rob Nixon has noted, "the past of slow violence is never past, [and] so too the post is never fully post."[2] Environmental justice requires openness to multiplicity. The Anthropocene names a pluralizing disorientation of our experience of time. How can we conceptualize the immediacy and also the apparent endlessness of an epoch that may have begun fifty years or ten thousand years ago but will clearly extend into many thousand years of climatological aberration? What does time feel like in this new dispensation?

Between Shakespeare, Dickenson, and Borges, *now* proves a slippery fish. Reading the Renaissance dramatist, nineteenth-century poet, and early twentieth-century fiction writer together provides insight into the human difficulties of chronological experience. The imaginative spear these writers co-compose focuses itself into the oblique power of Dickinson's envelope poems, ephemeral art built from scraps and words. Reading these three writers together rebuilds *now* into dizzying plurality.

Now (Iago)

Stirring up trouble outside his boss's father-in-law's Venetian apartment, Iago imagines time as a sticky fluid, staining what it touches. *Now* in his dramaturgy forces absent things into pres-

2. Rob Nixon, *Slow Violence and the Environmentalism of the Poor* (Cambridge, Mass.: Harvard University Press, 2011), 8.

ence, so that the eloped couple's nuptial bed appears in the city streets. He drives the vision into the sleepy father's eyes:

> Even now, now, very now, an old black ram
> Is tupping your white ewe! Arise, arise,
> Awake the snorting citizens with the bell
> Or else the devil will make a grandsire of you (1.1.87–90)

The pun between the "ewe" that is Desdemona and the "you" that is Brabantio as grandfather to devilspawn emphasizes how Iago's immediacy entangles human self-conceptions. Brabantio, not less than the other "snorting citizens," cannot arise fast enough to stop *now.* The devil, one of many racist formulations Iago generates for his superior officer Othello, owns the future. He "will make" Branbantio's descendants. Within this *now,* alternatives are impossible. Like climate change, Iago occupies all possible futures.

The contaminated *now* of Iago's rant beneath Brabantio's window also defines time's horizontal expanse in the metafiction of Jorge Luis Borges. In his story "The Garden of the Forking Paths," a World War I murder mystery maps onto the discovery of an ancient work whose title replicates that of Borges's story. The English Sinologist Stephen Albert informs the story's narrator, Yu Tsun, that he owns a copy of the book *The Garden of the Forking Paths,* which comprises "an enormous riddle, or parable, whose theme is time" (27).[3] The forking paths are the work of the ancient Chinese master Ts'ui Pen, and by an improbable coincidence Yu Tsun, the narrator, is himself a descendent of Pen, as well as being a German spy who urgently needs to pass a secret message to his allies on the Continent. The nested coincidences represent time's massive plurality, its options and impossibilities. Borges's richly ironic structure contrasts Ts'ui Pen's infinite

3. Jorge Luis Borges, "The Garden of Forking Paths," in *Labyrinths: Selected Stories and Other Writings,* trans. Donald Yates (New York: New Directions, 1964), 19–29.

labyrinth of time with the narrator's monomaniac violence. "I fired with extreme caution," Yu Tsun declaims passively, "I swear [Albert's] death was instantaneous—a lightning stroke" (28–29). The horizontal breadth of an infinite maze contrasts against the instantaneous strike of the killer's bullet. The bullet also communicates vertically to the German Air Force, which, having been informed of the name of their target by the murder of a man named Albert, bombs the city while Yu Tsun languishes in prison for murder. When do the deaths happen? Yu Tsun awaits the gallows, Albert dies in a lightning instant, the bombs fell "yesterday" (29), the Great War grinds on—and when Borges published this story, in 1941, a second iteration of the Anglo-German conflict raged in Europe. All these *nows* are out of joint.

When (Desdemona)

At the speculative core of "The Garden of the Forking Paths" sits the subjective experience of temporal multiplicity. Twice in the tale the narrator describes a "swarming sensation" (28). When Albert reads aloud from Ts'ui Pen's magnum opus, the narrator feels an "invisible, intangible swarming" (27). Later, just before killing Albert, Yu Tsun feels the multiplicity again: "It seemed to me that the humid garden that surrounded the house was infinitely saturated with invisible persons. Those persons were Albert and I, secret, busy and multiform in other dimensions of time" (28). Even earlier Yu Tsun had hinted at this multitudinous sensation when walking toward Albert's house: "The afternoon was intimate, infinite" (23). A world-idea in which the endless equals the near: that chronological humidity and denseness typify Yu Tsun's experience, both as spy who believes his nemesis has ensnared him, and as reader of his ancestor's book. When will the inevitable happen? Everyone, not just Iago, knows the answer: *now*.

Shakespeare's *Othello* is infamous among literary critics for its "double time scheme" that implies that the events in the drama either cover some months, leaving time for the purported affair between Desdemona and Cassio, or barely last one full day, since the newlyweds' progress to their wedding night gets interrupted twice—in Venice by Iago's rant, and in Cyprus by Cassio's drunkenness—before playing itself out as murder in the marital bed.[4] The simplest response to the multiple misfirings of time appears in the words of the wife with the "demon" in her name. "When shall he come?" Desdemona pesters her husband after he has banished Cassio, "Tell me, Othello. I wonder in my soul / What you would ask me that I should deny / Or stand so mamm'ring on?" (3.3.67–69). Desdemona's *when* institutes human choices in place of the endless swarming closeness of Iago's *now*. Her specific choice of time opposes the forking gardens of Ts'ui Pen. Desdemona imagines that it's possible to choose a time and act accordingly. It seems a good plan, and perhaps a good model for taking environmental action today. But the burden of *when* proves insupportable in the play.

Never (Othello)

Nearly four hundred lines after Desdemona's hopeful *when,* after crashing through a long waterfall of a scene that takes our hero from loving husband to homicidal killer, Othello names the time that is right for him:

> Never, Iago. Like to the Pontic Sea
> Whose icy current and compulsive course
> Ne'er keeps retiring ebb but keeps due on
> To the Propontic and the Hellespont:

4. This chestnut of traditional criticism appears most prominently in A. C. Bradley, *Shakespearean Tragedy: Lectures on Hamlet, Othello, King Lear, Macbeth* (London: Macmillan, 1919), first published 1904.

Even so my bloody thoughts with violent pace
Shall ne'er look back, ne'er ebb to humble love
Till that a capable and wide revenge
Shallow them up (3.3.456–63)

Othello's tragic firmness draws its metaphoric force from his untypical example of a sea whose tide runs only one way. The impossibility of this ebb-less ocean marks the Moor as refusing to accept the world as it exists; he substitutes violence and revenge for human and geographic rhythms. Like a modern climate denier, he rages over the physical realities of his world. No ebb means neither time nor tide can sway him, leaving him trapped inside Iago's recursive *now.*[5]

The figure in Borges for Othello's bloody and linear focus is neither the murderous German spy nor his doomed Sinophile host but instead the Irish policeman who tracks Yu Tsun to the garden. Like Othello, "Madden was implacable. Or rather, he was obliged to be so" (19). In fact, Madden also resembles Othello in that his ethnicity makes him suspect: "An Irishman in the service of England, a man accused of laxity and perhaps of treason, how could he fail to seize and be thankful for [the] miraculous opportunity" (19–20) that exposing the German spy affords. Madden hovers around the outside of Borges's narrative; the fall of his footsteps entering the garden spurs Yu Tsun's bullets. Later Madden facilitates the narrator's transfer to the waiting gallows. In Borges's miniature narrative, which eschews both melodrama and the massive chronological sprawl of Ts'ui Pen's novel, *never* plays only a minor role. The project of the story centers on "the abysmal problem of time" (27), and it is exactly that problem that men of violence like Othello and Madden fail to perceive.

5. For my previous thoughts on this passage and the maritime meanings of this play, see "Keeping Watch: *Othello*" in *At the Bottom of Shakespeare's Ocean*, 19–32 (London: Bloomsbury, 2009).

Contrition (Yu Tsun)

In place of the heroic and rhetorically dense deaths that Shakespeare's tragedy stages for his audience, Borges's Yu Tsun waits for the noose not with fear or rage but with "innumerable contrition and weariness" (29). He does not resemble voluminous Othello or other tragic heroes who orate their own demises, though in a sense he occupies the strange already-dead position of Hamlet who, feeling the poison working in his veins, cries, "I am dead, Horatio" (5.2.317) while still speaking.[6] The prince of Denmark, like Yu Tsun, speaks from beyond life and thus beyond the debt mortality owes to chronology. The particular insights into time that Borges's story unfolds derive at least partially from this peculiar position: Yu Tsun has acted decisively, killing Albert and succeeding in his espionage mission, but he overstays his time. Having been granted a vision through the garden of the forking paths of the endless possibility of times-not-yet-begun, he ends mired in the exhaustion and regret of time-that-has-passed. The contrition of Yu Tsun, like the faint English pun of his name ("you soon"), fixes the limits of human temporal engagement. Time may open endlessly, but it closes with the hangman's noose or the barrel of a gun.

The Point of the Spear (Emily Dickinson)

The poetic solution to time's hostile enclosures isn't only finding the sea marks through which we must steer our fragile bark. Neither Iago's angry *now,* Desdemona's forlorn *when,* Othello's enraged *never,* nor even Yu Tsun's melancholy *con-*

6. William Shakespeare, *Hamlet,* ed. Ann Thompson and Neil Taylor (London: Bloomsbury/Arden3, 2006).

trition solves the Anthropocene's time riddle. But there is a way through time, at the point a poet's spear.

More perhaps than any Anglophone poet, Dickinson matches environmental disorder with her own partial, individual, and flexible order. Human language, especially the semiprivate idiosyncratic language of Dickinson's poems, measures out a fractured response to the nonhuman dynamism we have recently begun naming Anthropocene. Against global catastrophe and the tyrannous rage of the Old Man, Dickinson's "envelope poems," written on oddly-shaped scraps of paper that were found in her desk alongside her manuscript fascicles after her death, assert fragmentary claims. One in particular, written in increasingly shortened lines leading down to the envelope fragment's triangular point, isolates the temporary, bounded, language-given force of human actions operating inside nonhuman constraints.[7] Enclosed by time, mortality, and a physical page that grows more and more narrow, Dickinson announces and defers her authority:

In this short life
that only lasts an hour
How much – how
little – is
within our
power

That "much" and "little" represent the intertwined forces of destructive Anthropos and redemptive Humanity, except that the Human isn't only redemptive and Anthropos not only destructive. Neither controls the whole poem or whole Anthropocene world. Instead the poet's final word, wedged into the point of

7. Identified in the Amherst archive as manuscript #252, this scrap can be seen on the Emily Dickinson Archive website: http://www.edickinson.org/editions/1/image_sets/238648. Accessed July 19, 2018.

the envelope spear that would have been aimed directly at the seated writer's stomach, reads "power." That human and post-human power enables the writing of new narratives out of old poems in our environmentally unstable present.

Errant Nature

LIKE A NOVICE THIRD BASEMAN, I feel the errors piling up around me. I'll make a few stabs at them here, remembering that erring isn't an orderly process. A good thing too!

Crafting an eco-language for postsustainability in a pluralized Anthropocene, I tend toward error. Since disruption and change are basic ecological principles, error—in the sense of turning, changing, surprising—represents a basic truth of the more-than-human composite we still mostly call Nature. Getting wet because I'm standing out in the atomic rain with Lucretius, I'm looking for plural ways to conceptualize ecological change.

Errors wears many faces. Philosophical error, legal error, the sinner's turn away from grace. Errors crop up in engineering, in grammar, logic, ethics, mathematics, baseball. To err is to wander or deviate, and from that unexpected turning possibilities appear.

Unpacking the depths of my personal fascination with errancy might require delving into the Little League of my childhood subconscious, but as an ecocritic my error fixation begins with feeling lost at sea. Early modern oceanic navigation forced sailors, writers, and poets into reckoning with nearly unsolvable errancy. I've got a bumper-sticker version of that insight: "The

Age of Discovery was an Age of Error."[1] In navigational context, error means arriving unexpectedly at a place that's not the one you were trying to reach. Errancy means you reach Cuba when you're sailing for China or wreck on the Scilly Islands when sailing home to Plymouth. Global and oceanic errors accompanied early modern sailors as they ventured into poorly known seas. Error was every voyage's shipmate.

Entangled with this mathematical and geophysical sense of error, which motivated a series of technical fixes from the Mercator projection to John Harrison's maritime chronometers, lurked the premodern obsession with Original Sin, a basic human turning away from divine law. To err is human, as the saying goes, but not only in a harmless way.

Being born into error may acquire new resonance as long-established ideas of human sinfulness transmute themselves into species-level guilt over having fouled our planetary environment. As we come to recognize the Anthropocene not as cliff to be avoided (too late for that!) but as the condition of our present and future, literary figurations of "endlesse worke" become newly valuable. From the knights and ladies who roam the labyrinthine allegories of Edmund Spenser's *Faerie Queene* (1590–94) to the more practical but still interminable *Spiritual Exercises* (1522–24) of Ignatius of Loyola, a shared premodern vision emerges of human insufficiency and the need for ceaseless labor. Spenser's quests and Ignatius's prayers inform the predestinatory labyrinths of John Calvin, as well as his Anglophone literary heirs including Herman Melville and Thomas Pynchon. The shock of living in Anthropocene error marks John Milton's brutal enjambing revision of classical myth in the first book of *Paradise Lost*:

1. Steve Mentz, "Mapping Uncertainty: Marine Cartography, the Wright-Molyneux Map, and *Twelfth Night*," *English Language Notes* 52, no. 2, "Cartographies of Dissent" (2014): 53–59.

. . . thus they relate,
Erring (1.746–47)[2]

The Christian poet insists that all who wrote before him wrote in error—but he knows that he too errs, and that failing to stay on the straight path represents a basic human condition. The world without Anthropocene pollution appears only as Eden itself, the loss of which has never yet stopped erring. After some turns, you can't find your way back to the former way. Eventually, the idea of a single path starts to seem nebulous. Perhaps it's time for many routes?

On the third hand—how many hands is that? Error!—errancy may turn out for the best. That's the fond hope and deeply held fantasy of the ancient literary form we call romance, in which sudden turns become fortunate coincidences, as they are revealed over the fullness of the long narrative voyage. In my favorite genre joke, Northrop Frye defines the origins of classical romance through navigational error: "In Greek romance . . . the normal means of transportation is by shipwreck."[3] In Spenser's *Faerie Queene* Errour is a monster, half "like a serpent horribly displaide" (1.1.14.7) and the other half womanly. She frustrates interpretation and for a time immobilizes the Knight of Holinesse: "God helpe the man so wrapt in Errours endlesse traine" (1.1.18.9).[4] Trapped by Errour and in errancy, the knight requires faith to set him free. Spenser's creature represents romance error by definition—the opening enemy in Elizabethan England's greatest verse romance—but where is she taking our knight and our poem? Into her den, into immobility: a place where you can't err any more because you can't move.

2. John Milton, *Paradise Lost*, in *The Complete Poems and Major Prose*, ed. Merritt Y. Hughes (New York: Macmillan, 1957), 230.

3. Northrop Frye, *The Secular Scripture: A Study of the Structure of Romance* (Cambridge, Mass.: Harvard University Press, 1978), 4.

4. Edmund Spenser, *The Faerie Queene*, ed. A.C. Hamilton (London: Pearson, 2001), 35–36.

So what is error as ecological principle? Navigational dislocation, original sin, romance circuity: what links these disparate deviations? Is the essential narratability of error, its crucial work in making stories, related to its corrupting theological force? Can we err without catastrophe—or can Anthropocene error and its mounting catastrophes be reimagined not as a once-and-for-all break with an alienated paradise but as a way to recognize the shifting and violent contingencies in which we live?

Any attempt to solve such problems courts—yes, you guessed it—more errors. But despite the risk of adding one more turn to the many-forked idea tree, I'll propose ecology as a cognate language for error in the Anthropocene. Linking error to ecology helps emphasize the centrality of disruptive social and ecological change in both premodern and modern cultural moments. Thinking with error may not help us untie all ecological or political knots, but perhaps this practice may help us recognize the difference between pleasing fantasies and deep-down encounters with Nature as nonhuman presence.

Error is ecological because ecological systems operate through movement and difference as they change in time.

Ecology is errant because, as the "new" or dynamic ecologists have been arguing since the 1990s, there is no permanent stability or "sustainability" to be found in the natural world.[5]

Error, not stasis, typifies the natural world.

Human mechanisms for navigating error do not involve "correction" to an assumed norm so much as learning to accommodate change.

5. On dynamic ecology, see Daniel Botkin, *Discordant Harmonies: A New Ecology for the Twenty-first Century* (Oxford: Oxford University Press, 1992). For my literary responses to Botkin's ideas, see "After Sustainability," *PMLA* 127, no. 3 (May 2012): 586–92, and "Strange Weather in *King Lear*," *Shakespeare* 6, no. 2 (2010): 139–52.

Recognizing that Nature and Error are twins can change ecological thinking in the Anthropocene. I agree with Donna Haraway, Timothy Morton, Bruno Latour, and many others who have been arguing for some time that any concept of "Nature" that insists on being separate from "Culture" is a problem, not a solution.[6] But I also doubt the word *Nature* is going away, so my aim is to help renovate or reconfigure this plastic term, in part by returning to its rich literary and philosophical history. Nature has a deeper intellectual past than Ecology. How might a Nature that expands to include not just nonhumans but also chronologically distant figures as well as dynamic change and disruption as constitutive principles operate? What if Nature and Error are not opposites but mutually co-conspiring?

Living in Nature requires—and sometimes rewards—errancy. Sailors and poets have always known this problematic truth. The Anthropocene may teach error's painful lessons on a planetary scale.

Living in Error is Natural. Unfortunately.

Nature loves to hide, says Heraclitus.[7] Perhaps Error names the principle through which Nature secretes itself?

Put most simply: Nature errs. What might follow from this heretical ecological antiprinciple?

6. Morton provides perhaps the pithiest articulation of this idea in the title of *Ecology without Nature,* but it's a widely held idea in posthuman ecotheory.

7. On this gnomic phrase, see Pierre Hadot, *The Veil of Isis: An Essay on the History of the Idea of Nature,* trans. Michael Chase (Cambridge, Mass.: Harvard University Press, 2008).

The Neologismcene

THIS PARTICULAR TERMINOLOGICAL GAME seems to be about up, and it's no surprise that Old Man Anthropos has won again. I don't think we'll be using any word except Anthropocene to describe the ecological present anytime soon. More's the pity, perhaps, but the Age of Man seems here to stay. The dominance of this term may not only be a bad thing; the term "Anthropocene," despite its problems, can serve as insecure password opening into new ways of thinking the human. I crave many flavors and textures, including posthuman varieties. If many wanderers can fit through the Anthropocene door, our task should be to make room for them all.

But as environmental humanists embark on the necessary labors to #pluralizetheanthropocene, it seems worth documenting some alternatives. There are many other 'cenes in the salad. No list is likely to be complete, since there may never be a more neologism-filled moment in the environmental humanities than now. A tally of the preterite 'cenes that are even now being passed over in favor of the familiar postures of Old Man Anthropos will not by itself accomplish the necessary pluralization, but this list can provide a skeleton key to intellectual variety. These are some of the pluralities we can labor to recover while facing the onrushing tide of the Anthropocene.

The current incomplete count sits at twenty-four 'cenes, in alphabetical order, Agnoto- to Trump-.

Agnotocene: Derived from the term "agnotology" in sociology and the history of science, which studies "the production of zones of ignorance" (198), this jaw-breaker is one of the many alternative 'cenes suggested by Christophe Bonneuil and Jean-Baptiste Fressoz in their wide-ranging and brilliant book, *The Shock of the Anthropocene* (2016).[1]

Anglocene: In a side note to their chapter on the Thermocene, Bonneuil and Fressoz remark that another viable term would be "Anglocene," a name chosen to emphasize the outsized contributions of the United Kingdom and the United States to global carbon emissions. Or, as they put it in more politically charged terms: "The overwhelming share of the responsibility for climate change of the two hegemonic powers of the nineteenth (Great Britain) and twentieth (United States) centuries attests to the fundamental link between climate change and projects of world domination" (117).

Anthrobscene: Coined by media ecologist Jussi Parrika, this term, which appeared in 2015, emphasizes the obscenity of today. He doesn't mince words: "To call it 'anthrobscene' is just to emphasize what we knew but perhaps shied away from acting on: a horrific human-caused drive toward a sixth mass extinction of species."[2] Linking current fascination with data mining to "the sort of mining that we associate with the ground and its ungrounding" (56), Parrika, whose elaborations on these ideas appear in his subsequent book *A Geology of Media,* deftly explores our continued dependence on the material underground, including the scramble for the rare earth elements that power the "geological extracts" (37) that are our iPhones.[3]

Capitolocene: The preferred term of the eco-Marxist historian Jason W. Moore, this 'cene argues that the environmental villain

1. Christophe Bonneuil and Jean-Baptise Fressoz, *The Shock of the Anthropocene* (London: Verso, 2016).

2. Jussi Parrika, *The Anthrobscene* (Minneapolis: University of Minnesota Press, 2014), 6.

3. Jussi Parrika, *A Geology of Media* (Minneapolis: University of Minnesota Press, 2015).

is Capitalism, not Humanity or even Man write large. Moore's *Capitalism in the Web of Life* explores the increasing force and disruption driven by capitalist exploration of the natural world from roughly 1500 through the present.[4] Working in the tradition of "world-ecology" thinking, Moore defines our current crisis as the "end of Cheap Nature" (108) and the reckoning of a bill coming due after five hundred years of frontier exploitation.

Chthulucene: Donna Haraway's term, from *Staying with the Trouble,* asks for more-than-human alliances with "diverse earthwide tentacular powers and forces" (101).[5] She pointedly rejects the label "posthumanist," and she also claims not to be invoking Lovecraft's cosmic figure. Her Chthulu does not equal his Cthulhu: "note spelling difference" (101). She proposes the slogan "Make Kin not Babies!" (102). She also suggests that the Chthulucene may be aspirational, and our current era of transition may be best described by Kim Stanley Robinson's term "the Dithering," from his sci-fi novel *2312*.[6]

Econocene: Coined by the environmental economist Richard Norgaard in 2013, this term shifts the focus from Anthropos to Oikos—or, to put it colloquially, it's the economy, stupid.[7] According to Norgaard, the rapid growth of economic activity during the twentieth century has become the driver of climatic changes. He suggests that modern political and bureaucratic leadership must change the way it manages economic practices in order to come to terms with the new reality.

Homogenocene: I first discovered this term, which strikes me as one of the most sinister in the lexicon, in Charles Mann's brilliant work of popular history *1493: Uncovering the New*

4. Jason W. Moore, *Capitalism in the Web of Life: Ecology and the Accumulation of Capital* (London: Verso, 2015).

5. Donna Haraway, *Staying with the Trouble: Making Kin in the Chthulucene* (Durham, N.C.: Duke University Press, 2016).

6. Kim Stanley Robinson, *2312* (New York: Orbit Books, 2013).

7. Richard Norgaard, "The Econocene and the California Delta," *San Francisco Estuary and Watershed Science,* October 2013: https://escholarship.org/uc/item/4h98t2m0. Accessed July 11, 2018.

World that Columbus Created.[8] Mann cites among his scholarly sources M. J. Samways, in a 1999 article in the *Journal of Insect Conservation*. The Homogenocene presents a terrifying vision of a world in which all things in all places grow increasingly the same in physical, ecological, and even cultural terms. Homogeneity represents a form of ecological death. I worry about this one every time I travel to some remote part of the globe and see people wearing New York Yankees caps.

Jolyonocene: A late addition to the salad, this odd-sounding 'cene refers to a new generation of anti-Brexit UK activists, several of whom share the first name Jolyon, a now-fashionable antiquated way of spelling the name Julian that, according to geography professor Alex Baker of the University of Durham, "reeks of middle class indulgences in child naming and cultural superiority." The activation of this formerly content generation of "suddenly disenfranchised centrists" ushers in a "new, tragic, age of activism" that remains uncomfortable with older radical traditions.[9]

Manthropocene: Oxford economist Kate Raworth's call to arms in the *Guardian* in 2014 emphasizes the overwhelmingly male composition of the Working Group for the Anthropocene.[10] She suggests that gender bias might distort the group's methods and conclusions, and she further hints that a northern-hemisphere bias—"the Northropocene"?—may be at work. Such male-dominated groups also characterize the committee that chooses the Nobel Prize in Economics and similar awards.

Misanthropocene: The economist Raj Patel, who has recently collaborated with eco-Marxist Jason Moore on the book *History of the World in Seven Cheap Things,* also coined the term "mis-

8. Charles C. Mann, *1493: Uncovering the New World that Columbus Created* (New York: Knopf, 2011).

9. See https://anomiegeographie.wordpress.com/2017/03/27/the-jolyoncene/. Accessed July 11, 2018.

10. Kate Raworth, "Must the Anthropocene Be a Manthropocene?" *The Guardian,* October 20, 2014: https://www.theguardian.com/commentisfree/2014/oct/20/anthropocene-working-group-science-gender-bias. Accessed July 11, 2018.

anthropocene" as a way to imagine what happens if we do nothing.[11] "We're surrounded by catastrophic narratives of almost every political persuasion," he writes, "tales that allow us to sit and wait while humanity's End Times work themselves out. The Anthropocene can very easily become the Misanthropocene."[12] "At the very least," he continues, "we have been warned."

Naufragocene: I coined this one in *Shipwreck Modernity* to leverage shipwreck as metaphor and material reality for the age of catastrophic environmental change that emerged with ecological globalization during the early modern period.[13] Global maritime traffic during this period increased the perennial risk of travel by water, and the experience of early modern sailors expressed themselves in dialogue with ancient and then-contemporary literary narratives in which shipwreck represents divine revelation, mortal hubris, and the dilemma of humans who interpose their own fragile bodies in more-than-human environments.

Necrocene: Justin McBrien's notion of "the Necrocene—or 'New Death' . . . reframes the history of capitalism's expansion through the process of *becoming extinction.*"[14] By generalizing extinction as both cultural and biological process, McBrien argues that "the process of *Necrosis* is central" (117) and that the modern dilemma has at last "inverted" Benjamin's famous Angel of History: "now we hurl forward, ignorant of the past, eyes fixed on catastrophe upon catastrophe piling up ahead" (120).

11. Raj Patel and Jason W. Moore, *A History of the World in Seven Cheap Things: A Guide to Capitalism, Nature, and the Future of the Planet* (Berkeley: University of California Press, 2017).

12. Raj Patel, "The Misanthropocene," *Earth Island Journal* (Spring 2013): http://www.earthisland.org/journal/index.php/eij/article/misanthropocene/. Accessed July 11, 2018.

13. Steve Mentz, *Shipwreck Modernity: Ecologies of Globalization, 1550–1719* (Minneapolis: University of Minnesota Press, 2015).

14. Justin McBrien, "Accumulating Extinction: Planetary Catastrophism in the Necrocene," in *Anthropocene or Capitolocene? Nature, History, and the Crisis of Capitalism*, ed. Jason W. Moore, 116–37, (London: PM Press, 2016), 116.

Phagocene: Another 'cene from Bonneuil and Fressoz, the Phagocene puts consumerism and "disciplinary hedonism" (157) at the center of climate destruction. They diagnose modernity as "throw-away culture" (159), which they connect primarily to twentieth-century American mass production of consumer goods, especially the automobile and its cognate, the suburb.

Phronocene: One of Bonneuil and Fressoz's more paradoxical coinages, the Phronocene explores the longstanding awareness by Europe's central planners and early ecologists of environmental vulnerability. They conclude ruefully that "our ancestors destroyed environments in full awareness of what they were doing" (196). In this view, efforts to increase "environmental awareness" seem futile, because although such awareness has been plentiful in the historical record, it has not yet succeeded in slowing humanity's destruction of nonhuman systems.

Plantationocene: I first spotted this one on twitter via Tobias Menely, but it also appears in print in recent work by Donna Haraway and Anna Tsing. In Tsing's compelling formulation, "Plantations are machines of replication, ecologies devoted to the production of the same."[15] (See also the Homogenocene.) The Age of the Plantation reformulates the Capitolocene so that the slave plantation, rather than the factory, represents the dominant economic and ecological model of progress and disaster.

Planthropocene: A recent coinage of medievalist ecocritic Rob Barrett, for his work in progress on premodern engagements with vegetable life.

Polemocene: Bonneuil and Fressoz use this 'cene to emphasize a long history of political struggle motived by social justice and what Rob Nixon calls "the environmentalism of the poor."[16] Resistance to industrialization and "progress," they show, is as

15. Anna Tsing, lecture "Earth Stalked by Man," University of California, Santa Cruz. Cited in Catherine Ashcraft and Tamar Mayer, eds., *The Politics of Fresh Water: Access, Conflict, and Identity* (London: Routledge, 2016), 189.

16. Nixon, *Slow Violence.*

old as the industrial revolution, which suggests to them that historical and political resources remain available to continue this struggle today.

Sustainocene: As championed by Harvard chemist Daniel G. Nocera, this neologism proposes an era of "personalized energy" made possibly through compact photosynthesis devices.[17]

Symbiocene: I found this one via the artist Cathy Fitzgerald, who cites Glenn Albrecht, a retired professor of sustainability from Murdoch University in Western Australia who also coined the term "solastalgia" for the mental distress caused by environmental destruction.[18] As an alternative to the "ecocide of the Anthropocene," the symbiocene "emphasizes ideas and practices to enhance the mutual flourishing of all life."

Thalassocene: My other coinage from *Shipwreck Modernity,* I adapt this neologism by way of the "new thalassology" of the environmental historians of the premodern Mediterranean Peregrine Horden and Nicholas Purcell.[19] In my global rather than Med-centric sense, the Thalassocene writes human and environmental history through and on the World Ocean, whose currents, storms, and massive capacity as a carbon sink shapes cultures, climates, economies, and futures.

Thanatocene: Bonneuil and Fressoz's term for an Age of Death reads the twentieth century's signature contributions to climate catastrophe through deadly global wars and ecological devastation. They emphasize that the crucial and catastrophic "petrolization of Western societies" owes a powerful debt to, and is perhaps unthinkable without, the global mobilization of the Second World War (138).

17. https://www.youtube.com/watch?v=u92O8LSkezY. Accessed July 11, 2018.

18. Glenn Albrecht, "Exiting the Anthropocene and Entering the Symbiocene," *Minding Nature* 9, no. 2 (Spring 2016): https://www.humansandnature.org/exiting-the-anthropocene-and-entering-the-symbiocene. Accessed July 11, 2018.

19. Peregrine Horden and Nicholas Purcell, *The Corrupting Sea: A Study of Mediterranean History* (London: Routledge, 2000).

Technocene: Swedish ecologist Alf Hornborg diagnoses the uneven spread of technological expertise and circulation as a driving factor in today's ecological crisis.[20] He calls for a closer exploration of the mutual entanglement of social and scientific concerns in exploring the modern environment.

Thermocene: In Bonneuil and Fressoz's "political history of CO_2," the familiar hockey-stick climate curves get placed in the larger context of industrial modernity. Insisting that we must "denaturalize the history of energy" (107) requires that we also acknowledge that the history of energy is "political, military, and ideological" (107).

Trumpocene: When first putting this lexicon together at the end of 2016, I included this 'cene as a painful joke and exercise for the reader. The past two years have seen a predictable boom in thinking about the Age of Trump. Perhaps the sharpest and angriest elaboration of the ecological subtext of Trumpism has come from Kyle McGee, whose open-source book *Heathen Earth: Trumpism and Political Ecology* radicalizes our sense of the U.S. president's climate policies: "By way of geoengineered global warming, the climate itself can become the principal American weapon in the endless war on terror."[21] Or, to put it more starkly: "the real Death Star is already here, in our abundant fossil fuel extraction" (98). It falls to us now, McGee insists, to respond with the forces of community: "alliance, assembly, occupation, strike, protest, march, demonstration, above all *appearance*" (144).

Crafting adequate responses to the vast plurality of the 'cene salad comprises the quixotic task of this moment in the environmental humanities. Let's get started!

20. Alf Hornborg, "The Political Ecology of the Technocene," in *The Anthropocene and the Global Environmental Crisis: Rethinking Modernity in a New Epoch,* ed. Clive Hamilton and Francois Gemenne, 57–69 (New York: Routledge, 2015).

21. Kyle McGee, *Heathen Earth: Trumpism and Political Ecology* (Brooklyn: Punctum Books, 2017), 91.

Acting Human. Being Posthuman

THE LAST STORY features Jonah, the prophet who dives deepest, and also returns.

The dive down from *human perspective*:

What would it mean to "go to Nineveh, that great city, and cry against it" (1:2)?[1] To protest human evil is a human task, perhaps the most virtuous of human tasks. If we look around during the Anthropocene, we see the "great city" of petromodernity, against which eco-prophets should and do cry: "for their wickedness is come up before me" (1:2). But Jonah, like the rest of us, remains human. He shirks his task, avoids the hostile city, takes ship, falls into what Herman Melville's Jonah preacher calls "his hideous sleep."[2] We know this unconsciousness because we drowse in its grip. Climate change follows the prophet as "a mighty tempest in the sea" (1:4). The mariners—afraid, as all of us are afraid—cast Jonah into the maelstrom. The "great fish" (1:17) that swallows him has long

1. "Jonah," *King James Bible (The Authorized Version)*, ed. David Norton (New York: Penguin, 2006) 1209–11.

2. Herman Melville, *Moby-Dick, or The Whale*, ed. Hershel Parker and Harrison Hayford (New York: Norton, 2002), 51.

since been analogized to the monstrous whale, the largest mammal on earth. English poet and activist Heathcote Williams calls whales "Alien beings," "Into whose mouths fourteen people could be placed with headroom."[3] That's where the prophet goes, into that alien space. Into it, and down.

Poets also supply reasons to dive down. The Australian poet Peter Porter's 1973 book *Jonah,* made in collaboration with the painter Arthur Boyd, portrays the diving prophet descending with human impulses:

> But it's fun in Whaleland, childhood
> Comes back with spidery jokes, the blue-
> tongued lizard and praise for being good![4]

Does Jonah want to swim or drown? What's his human goal in vacating the crowded ship? What worlds float in storm-churned nonhuman waters?

The dive down from *posthuman perspective*:

"The storm was a metaphor," explains the poet, "the incandescent elements tossed the ship like a child playing with fingerpaints" (26). Happy inhuman play and color point outward toward posthuman exfoliations.

For Herman Melville, the descent beyond the human makes some things obvious. "But God is everywhere" (53), growls Father Mapple, the whaleman-preacher. He sermonizes past human knowledge:

> God came upon him in the whale, and swallowed him down to living gulfs of doom, and with swift slantings tore him along "into the midst of the seas" where the eddying depths sucked him ten

3. Heathcote Williams, *Whale Nation* (New York: Harmony Books, 1988), 18, 22.

4. Arthur Boyd and Peter Porter, *Jonah* (London: Secker & Warburg, 1973), 39.

> thousand fathoms down, and "the weeds wrapped around his head," and all the watery world of woe bowled over him. (53)

Down at the bottom Jonah sees and also can't see. The descending prophet is multiply contained: his body inside the whale, his skull wrapped with weeds, the whale-human assemblage deep in the violet-black benthic zone ten thousand fathoms down. No one has gone closer to the real center than Jonah, "down to the bottom of the mountains" (2:6). In that darkness, Jonah prays but can't move. Alien pressures hold him still. He experiences but gains no knowledge. Dreaming of release, he makes promises—"I will sacrifice unto thee with the voice of thanksgiving" (2:9)—but he cannot act. Only the great fish acts, "and it vomited Jonah upon the dry land" (2:10). The human prophet returns, puke-covered and salt-stained. The great city won't know what hit it.

In declaring posthuman allegiance, the critic Rosi Braidotti notes that she "came of age intellectually and politically during the turbulent years after the Second World War," which brought her into contact with "feminism, de-colonization and anti-racism, anti-nuclear and pacifist movements."[5] Her anti-humanist politics rejects the masculinity and monomania of Old Man Anthropos. When Braidotti seeks out "a non-dualistic understanding of nature-culture interaction" (3), she imagines the opportunity to "decide together what and who we [humans] are capable of becoming" (195). Another influential thinker, Cary Wolfe rejects the "posthuman" as such in favor of a "posthumanist . . . sense that . . . opposes the fantasies of disembodiment and autonomy, inherited from humanism."[6] Wolfe finds a positive project for posthumanism not only in Braidotti's liberatory politics but also in a theoretical engage-

5. Rosi Braidotti, *The Posthuman* (Cambridge: Polity Press, 2017), 17.

6. Cary Wolfe, *What Is Posthumanism?* (Minneapolis: University of Minnesota Press, 2010), xv.

ment between literary theory, animal studies, and disability studies. Wolfe's embrace of radically difficult moments of vision and failure leads him to attack the "distinction between human and animal" (98) that has often seemed fundamental to discourses of the human and humanism.

Neither Braidotti's dream of recombination nor Wolfe's skeptical deconfiguration of humanism reach Jonah deep within the whale. Silent, prayerful, he sits inside too much flesh. The posthuman, at bottom, must be a silent dream, or perhaps a dream of an impossible fecundity, a still-gestating pregnancy beneath the heart of the world.

The ascent from *human perspective*:

In the contemporary Icelandic novelist Sjón's *From the Mouth of the Whale,* the experience of returning from the beast's throat appears a perfectly rational process. Sjón's hero Jónas the Learned "knows the species . . . he has been consumed by a north whale; an evil leviathan that grows to eighty or ninety ells long and the same in width, and its food is by all accounts darkness and rain, though some say it also feeds on the northern lights."[7] Human learning interprets the whale as almost allegorically beyond-human, feeding alternately on darkness and light. But the novelist like the poet sees the world for itself. In Porter's phrase, "The sky has two sides and a whale is only a symbol" (43). Inside that fleshy cavern, Jónas chooses reason over lamentation: "he calculates that he has been lying unconscious in the fish's belly for three nights and two days" (230). Rather than being the passive recipient of God-vomit, the Icelandic hero peers through the vastness and "sets off at a run, racing over the slippery tongue as fast as his feeble legs can

7. Sjón, *From the Mouth of the Whale,* trans. Victoria Cribb (New York: Farrar, Straus and Giroux, 2011), 230.

carry him, out of the whale's mouth" (230). For Sjón, survival comes from human knowledge and patient timing, not divine rescue. The whalebeast remains unknowable, and in the novel's final line it "gives a splash of its tail, and disappears once more into the deep" (231). But Jónas has been inside and come back out, been down and back up. He returns to the human family.

In Melville's angrier fiction, Jonah ascends because he displays "the true and faithful repentance: not clamorous for pardon, but grateful for punishment" (52). Preaching to Nineveh, Jonah performs the task we want our activist eco-heroes from Bill McKibben to Alexandria Ocasio-Cortez to perform today: "To preach the Truth in the face of Falsehood" (53). Humans return from the posthuman encounter with visionary knowledge. That knowledge changes human lives. Nineveh repents in the face of Jonah's preaching, puts on the "sackcloth," decarbonizes its economy, and God turns his wrath away from the great city. But Jonah—the one who has been to the bottom—burns with anger. Blind to Nineveh's reforms, he isolates himself outside its walls. "I do well to be angry," he cries, "even unto death" (4:9). Humans who have seen posthuman depths have no patience for change, no belief in a shared future, no compassion for the "sixscore thousand persons that cannot discern between their right hand and their left hand" (4:11). Apocalypse would provide a narrative answer—Jonah, like Hollywood, loves apocalypse—but it doesn't come.

The ascent from *posthuman perspective*:

We need today what the poet glimpsed in 1973. Jonah is the necessary eco-angel of our earth:

> Now I am to be a prophet without honour,
> one proclaiming Ecological Disaster,
> the Apotheosis of Pollution, the End of Spaceship Earth . . . (89)

Where is the repentance that saves Nineveh? Melville's truth is harder: Father Mapple's "two-stranded lesson" (49) punishes the "pilot of the living God" (53). Mapple is one of many steering pilots in *Moby-Dick,* whose ranks include Ahab and Ishmael, Queequeg and Pip. Only Mapple's Jonah truth speaks from inside the tempest. "Woe to him," he thunders, "who seeks to pour oil on the waters when God has brewed them into a gale!" (53). He chooses tempest, believes in storm, listens for the thunder. Rather than rescue, he craves woe upon woe:

> Woe to him whose good name is more to him than goodness! Woe to him who, in this this world, courts not dishonor! Woe to him who would not be true, even though to be false were salvation! (53)

The weeds twist around Mapple's logic, so that it becomes uncertain toward what port he pilots his living craft. Does he seek truth or salvation? How can these destinations differ? Father Mapple stops speaking and "fell away from himself for a moment" (54). Like Jonah, he loses awareness and forgets his fellow humans. Mapple returns to language not through the need to proselytize Nineveh but though an oceanic feeling: "But oh! shipmates! At the starboard hand of every woe, there is sure delight, and higher the top of that delight than the bottom of the woe is deep" (54). A mariner among mariners, Mapple invokes the ship's allegorical geometry: "Is not the main-truck higher than the keelson is low?" (54). The sermon's delight returns, in oblique recursive fashion, to the core humanist dream of individual freedom: "Delight is to him—a far, far upward and inward delight—who against the proud gods and commodores of this earth, ever stands forth his own inexorable self" (54). Mapple like Ahab stands apart, among "the personified impersonal, a personality" (382). Like Ishmael he survives his alien storm. As God's pilot he rejects the humanist practices of repentance and reform. No environment waits for exhausted Mapple after his sermon; he "covered his face with his hands, and so remained, kneeling, till all of the people had departed, and he was left alone in the place" (54). Like Jonah storming forth from Nineveh, like Ishmael scouring the seas after the *Pequod*'s loss, Mapple has seen the nonhuman. Vision exhausts words.

How can we combine our *human* need to reform the great city with our awareness of the *posthuman* plurality that environs our bodies?

The *human* in the prophet preaches repentance, change, and survival.

The *posthuman* into which the prophet dives promises shock, disorientation, and possibilities we cannot contain.

"Why leave the sea?" asks Luce Irigary of her marine lover, Friedrich Nietzsche. "Are you truly afraid of falling back into man? Or into the sea?"[8]

8. Luce Irigaray, *Marine Lover (of Friedrich Nietzsche)*, trans. Gillian C. Gill (New York: Columbia University Press, 1991), 12, 13.

Acknowledgments

The riotous plurality of the environmental humanities, a community of scholars, thinkers, and activists that overflows with passion, imagination, and good cheer in the face of catastrophe, has inspired this book. I'm grateful to several organizations who have hosted me and these ideas, including the Conférence Universitaire de Suisse Occidentale and Université de Lausanne, George Washington University's Medieval and Early Modern Studies Institute, the Environmental Humanities Centre and English Department at Bristol University, the James Edwin Savage Annual Lecture at the University of Mississippi, the Arthur F. Kinney Center for Renaissance Studies at the University of Massachusetts, and the Edmund S. and Nathalie Rumowicz Lecture Series at the University of Rhode Island. I thank the Glasgow Review of Books and Stanford's Arcade web magazine for e-publishing early versions of some of these pages. Fathom-deep gratitude goes to Vanessa Daws, who painted the image in the preface.

I write toward plural futures I cannot know and dedicate this book to our plural unknowing.

Steve Mentz is professor of English at St. John's University in New York City. He is author of *Shipwreck Modernity: Ecologies of Globalization, 1550–1719* (Minnesota, 2015), *At the Bottom of Shakespeare's Ocean,* and *Romance for Sale in Early Modern England.*